THE SPACEFLIGHT / REVOLTUION

SCIENCE, CULTURE, AND SOCIETY: A WILEY-INTERSCIENCE SERIES

BERNARD BARBER, Editor

Syntony and Spark—The Origins of Radio

Hugh G. J. Aitken

Astronomy Transformed: The Emergence of Radio Astronomy in Britain

David O. Edge and Michael J. Mulkay

The Spaceflight Revolution: A Sociological Study

William Sims Bainbridge

THE SPACEFLIGHT REVOLUTION

A Sociological Study

WILLIAM SIMS BAINBRIDGE

University of Washington
Seattle

A Wiley-Interscience Publication

JOHN WILEY & SONS, *New York • London • Sydney • Toronto*

Library of Congress Cataloging in Publication Data:

Bainbridge, William Sims.
The spaceflight revolution.

(Science, culture, and society)
"A Wiley-Interscience publication."
Includes index.
Bibliography: p.
1. Astronautics—History. 2. Rocketry—History.
3. Science fiction—History and criticism.
I. Title.
TL788.5.B34 301.5 76-21349
ISBN 0-471-04306-0

Printed in the United States of America

10 9 8 7 6 5 4 3 2 1

FOREWORD

Nearly 50 years ago sociologists William F. Ogburn and S. C. Gilfillan gave us a new way of looking at the processes of technological development. In criticism of what was then the established "heroic" paradigm for understanding these processes, they offered the new view of technological development as slow, incremental, cumulative, impersonal, almost inevitable. During these past 50 years, a good deal of scholarship by sociologists and economists has shown the considerable validity and usefulness of Ogburn's and Gilfillan's view.

Now, Professor Bainbridge, who has been an engineer and is presently a sociologist, has given us an additional considerable advance in our understanding of the processes of technological development. Ogburn's and Gilfillan's view is satisfactory for "normal" processes of technological development, he says, but not for revolutionary developments. The latter are developments carried out by men fanatically commited to a new technological vision, men who persevere and finally win against massive social, political, and economic indifference and opposition. They are willing to proceed by indirection and evasion, indeed by any path, to further their cause.

Professor Bainbridge's illustrative case for the presentation of his exciting new ideas is spaceflight, an innovation we all know to be fascinating from moon walks, space probes, pulsars, quasars, and science fiction. In this book spaceflight is the vehicle for Professor Bainbridge's theoretical innovations. Readers will be delighted by the new explanatory paradigm and the detailed story of how the Spaceflight Revolution actually occurred in Germany, England, and the United States. Most of the facts of this story have been publicly available, but they had not yet been collected and analyzed as they are by Professor

Bainbridge. In addition, through "field trips" to spaceflight centers, questionnaire surveys, intensive interviews with principal actors, content analysis of magazines, and attendance at conventions of science fiction devotees, he has collected still more evidence valuable both as support for his theory and as an account of what some of our fellow citizens think is important.

Professor Bainbridge treats the enthusiasts who brought off the Spaceflight Revolution as members of a successful social movement. During its revolutionary periods in the seventeenth century, science, too, was supported by a social movement in the same way that spaceflight has been. That social movement centered on the Royal Society. Sprat's *History of the Royal Society* and Glanvill's *Plus Ultra* were two of its self-conscious and ardent pieces of propaganda, asserting the superiority of the "new philosophy" to the ancient and established ideas.

When in the future we seek to account for the sources of the technological change, as a consequence of Professor Bainbridge's book, we will have to take account of the fact that these sources may be "revolutionary" as well as "normal." Professor Bainbridge has added considerably to our understanding of technological change.

BERNARD BARBER

CONTENTS

TABLES AND ILLUSTRATIONS

TABLES

ILLUSTRATIONS

Chapter 1 / INTRODUCTION

Spaceflight has been achieved despite the world's indifference and without compelling economic, military, or scientific reasons for its accomplishment. Not the public will, but private fanaticism drove men to the moon. When Neil Armstrong called his "small step" on the lunar surface a "great leap for mankind," he spoke as the partisan member of a revolutionary social movement, eager to convert the unbelieving majority to his faith. This book is a sociological analysis of that movement and its astonishing success in bringing about the Spaceflight Revolution.

Revolution might seem to be too strong a word, but the scale and manner of the achievement demand powerful language. Approximately $100,000,000,000 has been spent on space technology; the exact figure is debatable, but the order of magnitude is not. I use the word *revolution* as a scientifically descriptive term, not a metaphor. The development of spaceflight could be a revolution in two ways: its consequences or its causes.

Revolutionary Consequences

Wernher von Braun declared that the first moon landing was "... equal in importance to that moment in evolution when aquatic life came crawling up on the land."[1] Carried away by the enthusiasm of the moment, Richard M. Nixon said of the flight of *Apollo 11,* "This is the greatest week in the history of the world since the Creation."[2] Wags noted that Nixon had counted the moon landing as more important than the birth of Christ or even his own election to the presidency.

More sober minds have also believed spaceflight might be a breakthrough of such importance that its effects would be revolutionary. M.I.T. political scientist Lincoln P. Bloomfield wrote on "the Space Revolution," calling it "one of the physical transformations in the way men have lived, thought, and acted."[3] J. G. Crowther wrote a piece for *New Scientist* in 1969, saying:

> The exploration of space is potentially of absolutely major importance. Its beginning marks the emergence of Man from the confines of the Earth, and can be compared with the original emergence of life from the primordial ocean; it sets a new stage for evolution.[4]

In the history of human technology several revolutions have already taken place—relatively brief periods of rapid development in which man made "a fundamental change in his attitude to external Nature."[5] As V. Gordon Childe explains, the Neolithic Revolution was marked by the development of agriculture. "It was only after the first revolution...that our species really began to multiply at all fast."[6] Today the world faces the prospect of an end to population growth, and the view has been argued that only the colonization of other planets could allow the human population to expand further and continue the trend begun with the Neolithic Revolution.

Freeman Dyson of the Institute for Advanced Study in Princeton has even suggested that the further development of spaceflight may partly undo the social consequences of the Neolithic Revolution. When agriculture first made large cities possible, men ceased to live in the small groups that Dyson believes are our natural and most culturally productive units.

> I believe the real future of man in space lies far away from planets, in isolated city-states floating in the void, perhaps attached to an inconspicuous asteroid or perhaps to a comet....the ultimate benefit of space travel to man will be to make it possible for him once again to live as he lived throughout prehistoric time, in isolated small units. Once again his human qualities of clannish loyalty and exclusiveness will serve a constructive role, instead of being the chief dangers to his survival.[7]

G. Harry Stine has argued that the development of spaceflight and the consequent "exploitation of the space environment" will have the impact of a Third Industrial Revolution. The First Industrial Revolution, familiar to us all, "introduced powered machinery to replace human and animal muscle power."[8] The Second Industrial Revolution is happening today and is marked by the introduction of electronic computers and control devices to replace the human brain. In the Third Industrial Revolution, man's world itself will be supplemented or replaced by the resources and environments of outer space.

If space will in fact be used as the site for industrial plants is still very doubtful, and whether raw materials can be brought back profitably from the moon or asteroids is not yet clear. But, if successful, this Third Industrial Revolution might have at least two important positive results: The removal of mining and manufacturing enterprises from our planet would improve the terrestrial environment, perhaps making Earth a vast residential park; and as spaceflight proponents have argued for nearly 50 years, expansion of man's industrial base into the universe would circumvent the limits to growth imposed by the natural boundaries of our world.

The Space Age is too young for us to know for sure what its consequences will be. Although there is much room for argument, spaceflight may indeed have revolutionary consequences. If we forget for a moment the tremendous cost of further advance and guess that many new scientific and technical breakthroughs will permit an endless expansion into space, a grand vision of a future Interplanetary Civilization can rise in our imaginations. Will the economic production of the other planets exceed that of Earth? Will the population of a flourishing Earth be but an infinitesimal fraction of the total human population of the galaxy? Will future archaeologists search the universe for that lost planet on which the species arose millions of years before? Perhaps these are vain science fiction dreams. Were the *Apollo 17* astronauts the last men who will ever visit the Moon? Were the Apollo missions a frivolous "moondoggle" better forgotten? Is there no future beyond this small planet?

In a way, these last questions suggest that spaceflight is a revolution, even if it has no ultimate positive results. Either spaceflight will be proven a successful revolution that opened the heavens to human use and habitation, or it will be proven an unsuccessful revolution that demonstrated in its failure the limits of technological advance. Considering the technological optimism of many industrialists and government leaders, to learn that the sky imprisons us and that we must give up all hope of endless growth would be a major change. If spaceflight does fail, its abandonment will represent a technological counterrevolution of great consequence, symbolizing the end of progress as it is understood today.

Revolutionary Causes

In *The Sources of Invention,* Jewkes, Sawers, and Stillerman classify a number of major inventions according to whether they were made by individual inventors or by teams working in research laboratories. They found that "some cases seem to defy classification"[9] and one of these anomalous cases was the long-range rocket:

> Early interest in the long-range rocket, both on the practical and theoretical side, came from individual inventors, and enthusiasm was maintained by amateurs; but it was not until the German military authorities took up the subject during the Second World War that reliable rockets were built.[10]

To Jewkes and his colleagues, the development of long-range rockets looked like a mixture of individual and group invention. Really, it was an example of something quite different—invention and technological development by a social movement.

The importance of social movements in generating social and ideological change has long been recognized. Social movements reasonably ought also to be able to generate change in other spheres of culture, such as technology. While Jewkes and his associates say that "the German military authorities took up" the rocket, we find that almost the opposite was true. The Spaceflight Movement caused the German military *to be taken up by* the rocket. Germany, like the Soviet Union and United States later on, was exploited by members of the social movement for its own purposes.

This spaceflight movement was carried on by an extremely small dedicated network of men and was sometimes led by single individuals. Unlike some political revolutionary movements, it did not draw on general support from the population at large. Because its success was often in doubt and because major events turned on the actions of individuals, the development of spaceflight was not rigidly determined by the general advance of technology. It was a revolution that need not have happened. It was a revolution based on romantic idealism, not on practical rational considerations. Yuri Gagarin, first man to orbit Earth, wrote shortly before his death:

> Man's breakthrough into outer space, like other great accomplishments of mankind, must not be regarded only in the light of daily interests and routine practices. If throughout history people had been guided only by the satisfaction of their daily needs, mankind would still be living in caves.[11]

The motives of the handful of men who constituted the Spaceflight Movement are difficult to define exactly, but that they were not simple economic ambitions is certain. In 1946 Arthur C. Clarke urged fellow members of the British Interplanetary Society

> [to] be honest with ourselves. Any "reasons" we may give for wanting to cross space are afterthoughts, excuses tacked on because we feel we ought, rationally, to have them. They are true but superfluous—except for the practical value they may have when we try to enlist the support of those who may not share our particular enthusiasm for astronautics yet can appreciate the benefits which it may bring, and the repercussions these will have upon the causes for which they, too, feel deeply.
>
> The urge to explore, to discover, to "follow knowledge like a sinking star," is a primary human impulse which needs and can receive no further justification than its own existence. The search for knowledge, said a modern Chinese philosopher, is a form of play. Very well: we want to play with spaceships.[12]

Perhaps the most difficult step for the Movement was its evolution from a loose collection of amateur clubs who wanted "to play with spaceships" to major government programs actually building spacecraft and launch vehicles. In 1950 the editors of the *Journal of the British Interplanetary Society* summarized the history of spaceflight and identified the turning point as the German missile program of World War II: "...one man was chiefly responsible for a sweeping advance which at once moved the whole subject into the realms of practical engineering. That advance was the A.4 (or V.2) rocket, and the man was Wernher von Braun."[13]

Jonathon Leonard, science editor of *Time* magazine, has called von Braun "something of a prophet and something of a mystic," comparing him with Saint Frances of Assisi and Peter the Hermit. Von Braun does not disagree: "Enthusiasm and faith are necessary ingredients of every great project. Prophets have always been laughed at, deplored, and opposed, but some prophets have proved to be following the true course of history."[14]

Von Braun did not *follow* the true course of history; he *made* it. In Chapter 4 we dismiss the idea that military necessity gave him the millions he spent in developing the V-2 rocket. That the impetus behind spaceflight came from the general public or from scientists also is not true.

Six months before the first *Sputnik,* a national sample of Americans responded to a pollster's question on how much they were aware of plans to "launch a space satellite, sometimes called a manmade moon."[15] There had been a good deal of publicity in the mass media for months, spreading the news that the United States intended to put up such a satellite with the Vanguard rocket. But 54% had never heard of this plan at all! Despite 30 years of activity by the Spaceflight Movement and 30 years of extensive science fiction publishing in the country, only 20% could mention "at least some nontechnical ideas about the purposes of satellites."[16]

Raymond A. Bauer has conducted polls of public attitudes toward spaceflight and summarized the results of polls conducted by other social scientists. He reports: "At no point have any poll data indicated strong general support for the space program."[17] In a poll by the Opinion Research Corporation in 1966, respondents were asked: "If some of the federal government programs had to be cut, which of the programs listed on this postcard would you cut first?" The space program was chosen by 48%. In 1962 and 1963 the same research organization gave respondents a list of 26 issues and asked which required urgent action. "Developing rockets that land an American astronaut on the moon," came in next to last, just ahead of "financial support for artists and art activities."[18]

Support for crash programs to develop manned spacecraft has never been great among scientists either. This fact did not hold the program back in the mid-1960s simply because scientists participated very little in making the great decisions of the decade. Dr. Philip Abelson, editor of *Science* and a respected scientist in his own right, conducted a poll among scientists not personally connected with NASA. Of 113 he asked, 110 were opposed to the Apollo program, and three were in favor.[19] In May 1963 a group of 25 Nobel Prize winners announced themselves opposed to a crash program, without totally opposing the idea of space exploration.[20]

In 1964 *Science* mailed a questionnaire on space to 2000 members of the American Association for the Advancement of Science. Replies were received from 1134, including 548 with Ph.D. degrees. Only 20% of the whole sample and a mere 16% of the Ph.D.s felt that "a reasonable objective would be a lunar landing by...1970," the target President Kennedy had set three years earlier. Sixty-two percent of the whole sample and 64% of the Ph.D.s disagreed that "the vital national interests of the United States require that a high priority be given to landing a man on the moon by 1970."[21]

If von Braun had not succeeded in building the V-2 rocket for Hitler, the Spaceflight Revolution would probably have failed. A later attempt might have worked, but it is not likely. A delay of even a few months in the deployment of the V-2 would have prevented von Braun from demonstrating it in combat and thereby impressing the Russians and Americans with the capabilities of the long-range liquid-fuel rocket. In the 1950s a delay of perhaps five years would have reduced the chances for success of the movement by a tremendous margin. The Spaceflight Movement was racing against the rate of development of other branches of advanced technology. For example, the first ICBMs were all very large liquid-fuel rockets. They had to be large because the atomic warheads they carried were big and heavy. By the late 1950s atomic engineers had learned to build their bombs more efficiently and produced warheads that needed only relatively small rockets to deliver them. The largest ICBMs made sense for only a short time, and at that time bomber aircraft or pilotless winged missiles could deliver the atomic explosives quite effectively. If the Spaceflight Movement had not been ready and able to sell huge liquid-fuel boosters to the military in the early 1950s, it would not have enjoyed a second chance.

Not only was the Spaceflight Revolution achieved against the opposition or indifference of the majority of people, but like many political revolutions, it cost a number of human lives. About 4000 American soldiers were killed in the American Revolution. A similar number of citizens of London, Norwich, and Antwerp were killed in the V-2 bombardments. In April 1945 advancing American troops discovered 8000 corpses at the concentration camp associated with the Nordhausen V-2 factory.[22] Both the American Revolution and the Spaceflight Revolution may have been highly desirable, but both were carried forth over the bodies of dead and injuried human beings.

With considerably irony, Young, Silcock, and Dunn have written of the language with which von Braun's colleague, German General Dornberger minimized the bloodshed caused by the V-2 rocket he had helped develop:

> In postwar speeches he liked to call his rockets "flying laboratories," including, no doubt, the one that fell on a block of flats in Stepney on March 27, 1945 killing 130 people. "Altogether," Dornberger told an American audience in 1963, "3745 of these flying laboratories were successfully launched between September 6, 1944 and March 27, 1945. Some 1115 fell on England, 2050 on targets on the continent."[23]

Wernher von Braun's success in exploiting military and government leaders was certainly not an expression of popular will channeled through a revolution. It might be compared to a *coup d'etat* or a *palace revolution* in which power is seized suddenly by a small band of conspirators. Like such poorly supported political rebellions, the Spaceflight Revolution required exact timing, good strategy, and great luck to succeed.

Revolutionary Technological Change

Thomas Kuhn has drawn a distinction between *normal science* and *revolutionary science*. Normal science "means research firmly based upon one or more past scientific achievements, achievements that some particular scientific community acknowledges as supplying the foundation for its further practice."[24] Normal science is research building upon and following a *paradigm,* a tradition of accepted practice and of the proper questions for study. It builds step by step upon previous work. "Scientific revolutions are...those noncumulative developmental episodes in which an older paradigm is replaced in whole or in part by an incompatible one."[25] Without necessarily subscribing to Kuhn's model of the development of science, we can apply a similar distinction to technological development.

Normal technological change happens gradually, in response to conventional market pressures and in rational financial exploitation of new technological possibilities made available by other technological and scientific advances. Progress in some fields such as medicine may seem independent of what we usually call market factors or economic considerations, but still it is made in rational pursuit of benefit for human beings. S. C. Gilfillan has described the evolution of technology as resembling a biological process, happening in a natural inevitable way through the accretion of little details. He asserts, "There is no indication that any individual's genius has been necessary to any invention that has had any importance. To the historian and social scientist the progress of invention appears impersonal."[26]

Gilfillan himself studied the development of sailing technology and other examples of gradual evolution. What he says may well be true of *normal technological change;* I argue that it is not true of *revolutionary technological change.* Two factors might be responsible for the apparent impersonality of invention. First, the physical world makes only some inventions possible; inventors, in a sense, *discover* what can be done and how it can be done within the limits set by Nature. Second, if invention serves economic motives and the needs and wishes of society are taken for granted, the individual's influence is overpowered by the encompassing social order. These assumptions are often made by economic and technological determinists. Summarizing a number of studies, James M. Utterback reports:

> Market factors appear to be the primary influence on innovation. From 60 to 80 percent of important innovations in a large number of fields have been made in response to market demands and needs. The remainder have originated in response to new scientific or technological advances and opportunities.[27]

Economist Jacob Schmookler argues that "invention is largely an economic activity which, like other economic activities, is pursued for gain."[28] He believes that both categories mentioned by Utterback must be analyzed in economic terms:

> ...even when the idea for the invention is suggested by scientific discovery, the commitment to make it is generally an investment decision. This implies that even in such cases the invention is not an automatic outgrowth of the discovery. It is made only after estimating its value in the context of the times.[29]

While stressing economic factors, Schmookler dismisses any "thoroughly deterministic theory of history" on the grounds that some revolutionary innovations may be so difficult, so unlikely, that the role of chance or of the individual is decisive.[30] However impersonal and rigidly determined many kinds of invention may be, a few special cases have a great and unpredictable impact on human history.

In his study of German weapons research in World War II, General Leslie Simon considers "research due to commercial competition" and concludes: "Competition fosters refinements and the development of minor improvements, but it falls short of the stimulus generally required to bring out improvements which result in a whole order of increased effectiveness."[31] The V-2 long-range liquid-fuel rocket is one of the examples that inspired Simon's observation. It was not the result of *normal technological change.*

Revolutionary technological change is brought about through social processes that operate outside the conventional market mechanisms. Some important cases may have such unusual or obscure causes that they appear to be utterly indeterminant and inexplicable. Almost by definition, conventional social mechanisms are embodied in standard societal institutions, and revolutions of all kinds take place outside and in opposition to them. Therefore, revolutionary technological change should often emanate from social movements, forms of social action that transcend the purposes of standard society and operate outside its structures.

The greatest strides toward space were taken under the irrational dictatorships of Hitler and Stalin, and the American program was not instigated by big business interests. No corporation president would ever consider investing in an enterprise that promised to serve conventional needs only slightly better than already existing means and would require an investment of tens of billions of dollars and a delay of three decades or more between the initial investment and the first returns. Yet this was the case for spaceflight.

Spaceflight might have been a natural stage in the development of aviation. Some spaceflight pioneers, such as Valier[32] and Sänger,[33] imagined that aircraft speeds would steadily increase, year by year, until orbital velocity had been achieved. In fact the maximum efficient speed of aircraft appears to be on the order of Mach 3, only one eighth of the speed required to achieve orbit. Certainly, this is not how spaceflight *was* achieved. Spaceflight might have come about as the natural development of military rocketry. The range and payload of military rockets could have increased gradually, year by year, until orbital velocity had been achieved, and the size could have slowly increased until men could have been carried. Actually, military rocketry advanced in the train of space development. Men like Goddard and the team around von Braun improved military rockets merely as a way of advancing space technology. Had the size of military rockets been increased in a normal way, it would have been done much more slowly and would have been stopped at the current level of the solid-fuel ICBM, Minuteman, which is almost useless for even the most modest space tasks.

As Kuhn saw it, the structure of a scientific revolution involved a radical shift in intellectual paradigm. Technological revolutions also involve radical shifts of at least two kinds. There can arise a totally new conception of what can be achieved or a strikingly new standard of what it is desirable to achieve. The Spaceflight Revolution is probably an example of both types. The multistage liquid-fuel rocket is certainly a new concept of how people and machinery can be transported! But, perhaps most important, the exploration and conquest of space demand a transformation of values. However many practical justifications may be devised to urge pedestrian people to invest in space, all the early pioneers in space technology described spaceflight as an end in itself, a new value that cannot be reduced to a mere means of satisfying conventional objectives. In its most perfect form, the ideology of the Movement states that it is the purpose of the human species to conquer or enrich the universe; thus, it urges a revolutionary leap in our efforts and aspirations.

An Outline of this Book

Both an overview of the whole Spaceflight Movement and a detailed analysis of several specific fragments of it are of it are presented in this book. In the first four chapters the historical development of the movement and its significance are examined, with stress on the German wing. The last four chapters are a look at some American and British parts of the movement, and its future course is suggested.

The origin of the movement and the individual men who gave birth to it are described in chapter 2. In the third chapter we analyze the history of the crucial German branch, interpreted through a simple model of strategic interaction. We explain how the elite members of the movement were able to exploit Nazi Germany, the Soviet Union, and the United States to achieve spaceflight. In Chapter 4 a technical analysis of the accomplishments of the Germans during World War II is presented, using technology assessment both to convey the nature of the developments themselves and to tie them into a larger social context.

In the fifth chapter we scrutinize and compare the two oldest existing space travel societies, The American Interplanetary Society and The British Interplanetary Society. We focus particularly on their development from tiny amateur clubs, through processes of growth and institutionalization, to their current status as established scientific societies. Chapter 6 is a comparison of the very different Committee for the Future, which began as an amateur group dedicated to the colonization of the solar system, then envolved into a rather mystical, almost religious organization with no interest in space technology. The history and nature of the science fiction subculture is charted in Chapter 7, showing that it was important in the early days of the Spaceflight Movement but became ghettoized and unimportant later on. The future of the movement is suggested in the last chapter through an extrapolation of the social movement model to take into account the most recent technical developments.

Much of the data used in this study are taken from readily available published sources. How much information is available and how little social analysis has been performed using it is astonishing! Most authors who have written about the history of spaceflight felt that to string well known facts together into unimaginative chronicles was sufficient. The only books or articles in which any kind of social scientific analysis is attempted are those on the narrow historical period from *Sputnik I* to *Apollo 11*. With the exception of a few political scientists, social scientists have ignored spaceflight as a topic for study. However, many quasi-sociological judgments litter various books and articles by members of the Movement itself, most of them embarrassingly untouched by any understanding of the social sciences. This book is certainly not the last word on the subject of spaceflight. On the contrary, I hope it will mark the beginning of true sociological research in this important area.

The chapters on the Committee for the Future and the science fiction subculture are based in great part on ethnographic field research. I attended two conventions of the CFF, in Carbondale, Illinois, and Washington, D.C. I participated in six science fiction conventions—three regional "cons" held in Boston and three world science fiction conventions, held in Boston, Los Angeles, and Washington. I conducted many interviews, took hundreds of photographs, performed observations, and collected literature at all these meetings.

I also went into "the field" to get some personal perspective on the movement and its technology. I visited Cape Kennedy (now Cape Canaveral again); Huntsville, Alabama; and the area around Vandenberg Air Force Base in California. Although I did collect data on these excursions, the chief purpose was to get a global and somewhat intuitive feel for the concrete realities of the movement. Along with the Smithsonian in Washington, these places have extensive collections of launch vehicles and spacecraft.

Further data were collected through six questionnaires—five mailed to appropriate respondent samples, the sixth administered by my students at a science fiction convention. The questionnaires were designed to examine a variety of very specific issues. Although they provided me with an even better background for writing this book, much of the data gathered through them are not reported here. For example, I have extensive data on the ideology of the Spaceflight Movement, which may ultimately be published in a journal article. Sociologist Murray Dalziel and I are currently preparing an article, based on my questionnaire data, in which we analyze science fiction literature through quantitative techniques. The questionnaires and their respondents are described in Table 1.1.

At some point the reader may ask if this book is in favor of spaceflight or against it? I have no clear answer. The dream of space travel is glorious; the contemporary reality is dismal. Many men have been captivated by the dream in childhood and have grown up to be proud engineers, effective members of society; but they are forever earthbound. For the rest of us, space is a marginal field of activity of the military-industrial complex. Astronomy has flourished in recent years. That is a "good thing" but as much the result of conventional observation techniques as of space probes.

The success of the Spaceflight Movement is bittersweet. We have touched the moon and brought back information from Jupiter. By the end of this century we will have vast quantities of data about the rest of the solar system, but the original purposes of the Spaceflight Revolu-

TABLE 1.1 THE SIX QUESTIONNAIRES ADMINISTERED FOR THIS STUDY

Name	*Year*	*Respondents*	*Replies*
Q-NESFA-73	1973	Members of the New England Science Fiction Association	74
Q-NESFA-74	1974	Members of the New England Science Fiction Association	81
Q-FANZ	1974	Editors of Science Fiction amateur publications and their associates	130
Q-BOSK	1975	Participants in the Boskone Regional SF Conventional	79
Q-CFF	1973	Participants at the Syncon Convention held in Washington, D.C. by the Committee for the Future	80
Q-AIAA	1974	Probability sample of the 1973 membership of the American Institute of Aeronautics and Astronautics	102

tion had almost nothing to do with astronomy. The great pioneers of spaceflight wanted to perfect the spaceship so we could go to the planets, plant outposts and even colonies, and perhaps explore the stars. The promise of space has not been fulfilled, and at the close of the second decade of the Space Age, we do not yet know if it ever will be.

Chapter 2 / THE PIONEERS OF SPACEFLIGHT

Space historians frequently refer to one or another pioneer as belonging to the first or second "generation" of space scientists and engineers. I categorize creative intellectuals working in this field into four generational groupings:

GENERATION 0: Intellectuals who devoted energy to the question of spaceflight *before* technology had advanced to the point at which the right conclusions could be reached. They may be prophets with honor, but their work was probably not of historical significance. Examples: Kibal'chich, Ganswindt.

GENERATION 1: Intellectuals who laid the correct theoretical groundwork for spaceflight. Because of their own talents and the state of science and technology, they were able to come to the correct conclusions. In the world history of space travel, there were three such men: Tsiolkovsky, Goddard, Oberth.

GENERATION 2: Dynamic men who were able to gain influence over considerable resources and over other men to produce actual spacecraft after the sketches of the previous generation. They lived in historical circumstances that permitted them to exploit social conflicts to this end and were able to assemble large programs and expend billions of dollars. The two most important examples are von Braun and Korolyov.

GENERATION N: Intellectuals who perform useful innovative work within an already institutionalized Spaceflight Establishment. They come too late to participate in the creation of the Movement itself, but are able to create and invent within the context of the work of previous generations. As Generation 0 has no specific beginning, Generation N has no end.

The distinction I draw between Generations 1 and 2 may represent a basic difference between men engaged in producing theory and those engaged in praxis. Generation 1 might be described as scientists, Generation 2 as engineers. Generation 1 acted in social isolation, while Generation 2 built huge social enterprises. Neil Smelser suggests, "...we may distinguish between two kinds of leadership—leadership in formulating the beliefs and leadership in mobilizing participants for action."[1] Generation 1 performed the first function, Generation 2, the second.

In current language, Generation 1 engages in invention, Generation 2 in innovation. In a recent study Edward B. Roberts finds that "the difference between mere technical invention and successful innovation is largely attributable to the personal role of the entrepreneur"[2] and argues in favor of the "theory of the product champion as essential to technical innovation."[3] Wernher von Braun wrote in about 1950 that a mere inventor could not achieve spaceflight. "Since the actual development of the long-range liquid rocket, it has been apparent that true space travel cannot be attained by any backyard inventor, no matter how ingenious he might be."[4]

Von Braun himself was the most important technological entrepreneur in the German Space Program. In the histories of all major spaceflight efforts we find such men building rockets and institutions on the ideas of Generation 1.

We would expect the men of the earlier generations to be *more deviant* than those of the later generations. They were outsiders, while the later men are insiders. Correspondingly, we might expect each later generation to have a markedly more conventional lifestyle and intellectual outlook. This evolution of the membership from deviant to conventional should parallel the evolution of the Movement from collective behavior to societal institution, and each specific organization within the movement should show the same evolution. In the following pages I begin by examining the characteristics of members of Generations 0, 1, and 2.

Generation 0: Kibal'chich and Ganswindt

Intellectuals of Generation 0, almost by definition, appear to contemporaries as quacks, cranks, pseudoscientists, and madmen. Their reach exceeds their grasp by a wide margin. Martin Gardner suggests that the pseudoscientist is liable to display two characteristics: isolation and alienation from the traditional social organization of science and "a tendency toward paranoia."[5] Lemert has shown that "paranoia" is often the result of the social exclusion forced on a deviant who comes rightly to see other people as opposed to him,[6] and the entire Labeling Theory approach to a variety of mental disorders lays stress on societal reactions to men who are, simply, deviant.[7]

Soviet accounts of the life of Nikolai Ivanovich Kibal'chich portray him as a man alienated from conventional society against his will. Innocent of any wrongdoing, it is reported, he was sentenced in 1875 by the Czarist government to 2 years 8 months in prison for the mere possession of revolutionary literature which had been left in his apartment by a girl friend. Rynin says, "Prison and terrorism by the government turned Nikolai Ivanovich into a revolutionary."[8] He became an active member of a small revolutionary party and helped to prepare explosive devices for a number of terrorist bombings. After the Czar was assassinated by one of his infernal machines, he was arrested and executed in 1881.

While in prison awaiting trial and his inevitable execution, Kibal'chich wrote a brief article describing in general terms a rocket-propelled aircraft that would use gunpowder as its fuel. This monograph is broadly conceptual, rather than a design document; and because Kibal'chich had no access to a library, it contained little quantitative material. The essay was presented to the Czarist officials with the request that it be submitted to a panel of experts for evaluation and comment. The Imperial Police commented that "now is hardly the time to send this to experts for evaluation since it may only cause undesirable talk."[9] The article was buried in the police files until 1917, when it was unearthed and printed, rather too late to have any technical influence on the development of rocketry or progress of aviation.

Ley reports that Kibal'chich appeared uninterested in the course of his trial, but took every opportunity to engage in animated discussions of the technical details of explosive devices or his rocket aircraft idea. The picture conveyed by Ley's and Rynin's accounts is that of a monomaniac, unconcerned with the normal issues of life or death.[10]

Sergei Korolyov, Chief Designer in the Soviet space program, is said to have read Kibal'chich's article in 1918. He was greatly impressed by the spirit of the man and at least pretended to respect Kibal'chich's spaceship theory.[11]

Kibal'chich's vehicle was *not* intended as a spaceship; it was, as he called it, "an aeronautical machine."[12] Although the principle by which it would be propelled was that of the rocket, Kibal'chich discussed rockets only as a way of explaining the principle and said; "The example of the rocket is not altogether applicable here, since rockets have flight speeds which would be useless for an aeronautical machine."[13]

Rockets went too fast for his purposes! In fact, he discusses the flight of his craft by explaining first how it might hover in one place, and he never touches on the question of going beyond the atmosphere in it. Many writers had previously discussed the application of jets of steam or other gases as means of propulsion for aircraft; Butler and Edwards received a British patent for a rocket aircraft design in 1867.[14] Why, then, is Kibal'chich revered as a "prophet with some honor" in the history of spaceflight?[15]

Robert K. Merton has noted "the great frequency with which the history of science is punctuated by disputes, often by sordid disputes, over priority of discovery."[16] Generations 0 and 1 of the Spaceflight Movement have been parties to such disputes, willingly or unwillingly, during their lifetimes or posthumously. The *Prioritätsstreit*[17] over who should get the honor for inventing spaceflight has a certain nationalistic tinge. Kibal'chich was the earliest *Russian* who might reasonably be called "father of spaceflight"; Ganswindt was the first *German*.

Kibal'chich was also a revolutionary, and the bare facts of his life make him an excellent cultural hero for the Soviet government. Furthermore, the Spaceflight Movement itself has always had a strong historical interest, and proponents labored in the early years with considerable energy to construct a past and a tradition upon which it could build. Characters like Kibal'chich were appropriated to this emerging history, however irrelevant their work actually was. Finally, almost all histories of space development are mere popular journalistic stories, with no interest in social analysis, but rather a love of the striking and the colorful.

Hermann Ganswindt, the most mentioned German "prophet with some honor," was a clear candidate for the appellations *crank, paranoid,* and *pseudoscientist.* Son of a wealthy family, headed for a grand career in law, he dropped out of the University of Berlin to pursue his fantasies and excoriated the legal profession in a pamphlet ironically titled "The Last Judgment."[18] He began work on a dirigible airship and opened a small factory in which to develop his various inventions. He styled himself "The Original Inventor of Airships, Aeroplanes, Motor Cars, Internal Combustion Engines, Free-wheeling Systems, etc., etc."[19] Later in his career he claimed to have been the true inventor of the rocket-propelled spaceship. He publicly discussed the possibility of such a craft as early as 1891, but there is no proof of when he first came upon the idea.[20] In 1901 he was accused of fraud when it was claimed that a helicopter he had displayed did not fly under its own power, but was secretly raised up by a cable. An investigation determined that Ganswindt had neither achieved true flight with his craft nor was guilty of fraud. Hokum, however, was a major ingredient of his inventions.

Ganswindt's spaceship was an impossibility. Under the incorrect impression that a jet of gas could never provide as much propulsive force as a fusilade of solid bodies, his rocket accordingly was to eject 5-kilogram chunks of solid metal.[21] Ganswindt had no scientific training, and his spaceflight ideas were almost immediately debunked by better educated writers.[22] There is no evidence that his work had any positive influence on later rocket pioneers.

Ganswindt never had a political party acting as his historical advocate but was an active publicist on his own behalf. In 1925, when Oberth's work was beginning to be known, he attempted again to assert his own priority, without much success.[23] The fact that he was a German, and supposedly beat Tsiolkovsky in the race to invent the spaceship, did, however, encourage some German writers to spread his fame.[24]

We have little reason to believe that the Generation 0 men were of any historical significance. They neither inspired the Generation 1 workers, nor informed them of the correct route to follow. They are mere anecdotes, not antecedents.

Generation 1: Tsiolkovsky, Goddard, and Oberth

The three important Generation 1 men, Tsiolkovsky, Goddard, and Oberth, independently came to the same conclusions:

1. Spaceflight would be possible.
2. Of all known means of propulsion, only the rocket was suitable.

3. The liquid-fuel rocket engine, ideally using oxygen and hydrogen as propellants, would be the most powerful and efficient foreseeable type.
4. Although single-stage rockets could not escape the Earth, multi-stage vehicles capable of imparting to the final stage the necessary velocity could be constructed.
5. The basic formulae for engine performance and vehicle trajectory could be specified.
6. Various technical details and solutions to various practical problems of spaceflight could also be specified.

These discoveries represent a case of what Merton called a "multiple," the almost simultaneous or effectively independent invention of the same thing by different people. Oberth, who became aware of Goddard's earlier work just as he was about to publish his own first monograph, went to great pains to assert that his own work was original.[25] After the great public interest in rockets in the late 1920s, that a fourth great man would reinvent the same advances became unlikely, because the potential fourth discoverers would probably be aware of the earlier work. The six conclusions above represent a cluster of ideas "whose time had come," and which were almost bound to be discovered, even if our three *great men* had never lived.

However, these ideas, although crucial, did not in themselves yet constitute the invention of the spaceship. Tens of billions of dollars would have to be spent to develop real spacecraft from these rough notions. As T. S. Eliot said, "Between the idea and the reality ...Between the conception and the creation ... Falls the Shadow." However likely the ideas of spaceflight were, in the first third of this century the actual accomplishment of spaceflight was unlikely. Very special events turning on the deliberate actions of a few men were necessary to change the odds.

While there can be no doubt about Tsiolkovsky's theoretical accomplishments, we do not possess accurate information about his life and the social processes that influenced his work. Unfortunately, our information about all other Russian rocket pioneers is even less reliable. Russian defector Leonid Vladimirov has gone so far as to assert that Tsiolkovsky's work was of little significance, that he was resurrected from obscurity for Soviet propaganda purposes, and that various details of his biography have been rewritten to achieve the greatest effect. For example, Vladimirov claims that Tsiolkovsky was a Pole, not a Russian[26]

Tsiolkovsky's father was a sometime inventor and freethinker who never achieved the slightest hint of wordly success. "My father was a failure—an inventor and a philosopher."[27] Riabchikov notes that the father had been a forest ranger, but was dismissed from his job. "The real cause for the dismissal of the elder Tsiolkovsky was the fact that he—a man of rare honesty and straightforwardness—had never concealed his sharp hostility towards the tsarist regime."[28]

This is the Soviet account. For our purposes, we need not decide whether political reasons were behind the dismissal. In either case, Tsiolkovsky grew up in a home environment that included not only intellectual stimulation but *ressentiment.* We can speculate that his father's intellectual creativity afforded some compensation for an abysmally low social and economic status and that this pattern of response to misfortune was learned by the younger Tsiolkovsky. When he was eight years old his mother gave him a toy hydrogen-filled balloon.[29] Konstantin Eduardovich Tsiolkovsky, in this year of 1865 or 66, was particularly impressed by this toy. He writes:

> It also seems to me, though I could be wrong, that the basic drive to reach out for the sun, to shed the bonds of gravity, has been with me ever since my infancy. Anyway I distinctly recall that my favorite dream in very early childhood, before I ever read books, was a dim consciousness of a realm devoid of gravity where one could move unhampered anywhere, freer than a bird in flight. What gave rise to these yearnings I cannot say; I never came across stories like that, but I dimly perceived and longed after such a place unfettered by gravitation.[30]

> I was passionately fond of reading and devoured everything I could get hold of. I loved to indulge in daydreams and even paid my younger brother for listening to my recitals of them....For instance, I would dream that, since my brother and I were small, everything around us—houses, people, animals—were small, too. Then I would dream of being exceedingly strong. In my imagination I could jump higher than anybody else, climbed poles like a cat and walked ropes. I dreamed that there was no such thing as gravity.[31]

This theme appears again and again in Tsiolkovsky's writings, and of course in free space there is no sense of gravity. The conceptualization of a gravity-free space makes easier a number of the basic calculations of rocket behavior. In a humorous essay, "The Gravity Hater," he parodies himself.

> One of my friends was a very odd fellow. He hated terrestrial gravity as though it were something living; he hated it not as a harmful phenomenon, but as his personal, bitterest enemy. He delivered threatening, abusive speeches about it and convincingly, so he imagined, set out to prove its entire worthlessness and the bliss that "would come to pass" through its abolition.[32]

This "gravity" is not so much physical as political, for the Gravity Hater. "You have to thank gravity ... for pressing you down to the Earth like worms, for shackling you...." "But an environment free from gravity ... that's the thing! It makes the poor equal to the rich...."[33] Riabchikov, presenting the Soviet line, specifically analyzes Tsiolkovsky's attitude toward gravity as a metaphor for political oppression: "...in his dreams of a life 'without weight,' Tsiolkovsky also had another kind of weight in mind. He was also thinking of other chains—the chains that fettered the life of the people."[34]

I am pessimistic about the possibility of conducting a psychoanalysis-at-a-distance of any of the rocket pioneers. However, we must note that Tsiolkovsky's writings about spaceflight sometimes express a deep rapture, connected frequently to weightlessness, a sensuousness that suggests the ideas had deep, presocial significance for him. The mystical or fusion experience is described by a number of authors as involving a sense of weightlessness, like that which we may imagine is felt by the unborn child in the womb.[35] As it happens, sadly, real astronauts and cosmonauts report that weightlessness does not produce rapture.

While Tsiolkovsky must have been aware of his father's and family's status when he was a child, and might have infused his dreams with an inarticulate prepolitical analysis, his youthful fantasies more likely express his personal feelings of powerlessness and longing for freedom and strength. In these early fantasies *he personally,* or fictional protagonists who are clearly projections of himself, surmount gravity, not his father nor some mass of oppressed proletarians and peasants. The hero of "On the Moon" awakens unexpectedly in the weak gravity of the Earth's satellite, and wonders, "Have I grown stronger?"[36] By their very natures, a small child is weak; a grown man, by comparison, is strong. A child can compensate for his powerlessness, both his physical weakness and lack of social influence, through fantasies of being powerful. Science fiction, traditionally stories written for boys, exploited the longings of its young audience and fed their power fantasies. For the normal child in an easy environment, these unformed longings can be channeled through conventional activities by the subtle pressures of his social group. But illness cut Tsiolkovsky off from the normal course of social development. When he was nine an attack of scarlet fever ruined his hearing. He had difficulty communicating with children his age. He compensated for this deficiency by dreaming up heroic fantasies, increasing still further the gap between him and other people.[37] His life, consequently, was "poor in human contacts." Oddly, Tsiolkovsky's biographies and autobiographical statements say very little about his wife and children. Social isolation had become a motif, a way he saw himself, whether he actually continued to be isolated or not.

Tsiolkovsky was marginal to the scientific fields in which he worked. He was almost entirely self-taught and often ignorant of the prior and publicized work of other men. At one time he devoted his energies to developing a perpetual-motion machine.[38] At the age of 23, he submitted some papers to the St. Petersburg Society for Physics and Chemistry. The readers were astonished because his work was both brilliant and outdated. Tsiolkovsky had rediscovered several important scientific principles years after they had been publicized in the West.[39]

His isolation deprived Tsiolkovsky of scientific knowledge, but also freed him from the mental constraints and research agenda of conventional scientific thought. He was undeterred from considering with a fresh mind questions ignored by other scientists. Many social theorists have suggested that the man most likely to come to new insights is not the man best trained in conventional wisdom. Simmel, for example, notes the special objectivity of the stranger.[40] Jewkes, Sawers, and Stillerman stress the importance of "the uncommitted mind" in achieving inventive breakthroughs:

> The essential feature of innovation is that the path to it is not known beforehand. The less, therefore, an inventor is precommitted in his speculations by training and tradition, the better the chance of his escaping from the grooves of accepted thought.[41]

Kibal'chich and Ganswindt certainly possessed uncommitted minds, but this was not enough. Gilfillan writes:

> The inventions which revolutionize a device or industry are commonly made by men *outsiders* to it yet informed regarding it; the far greater and more valuable mass of perfecting inventions are made by *insiders.*[42]

Gilfillan should not say *outsiders,* but *marginals.* As he recognizes, the complete outsider with no understanding of a field at all cannot contribute to it. What distinguishes Tsiolkovsky from Kibal'chich and Ganswindt is that he had trained himself in basic physics, and *his mathematics was correct.* He acquired the proper intellectual tools from books and correctly applied them. To examine this crucial fact and to draw the needed line between our Generation 1 men and pseudoscience, we consider a *false* invention, a false means of space travel, invented *by all three,* then rejected in favor of the correct means.

When he was 16 years old, Tsiolkovsky imagined he had solved the problem of the conquest of space. He had become "interested in the theory of the centrifugal force" in the belief that "it could be used to lift a spacecraft from the earth."[43] He devised a space drive based on this principle:

> The apparatus he invented consisted of a closed chamber [or box] within which vibrated two pendulums with balls attached to their upper ends. The balls described circular arcs in their movement, and, Tsiolkovsky thought, their centrifugal force would lift the box and carry it into interplanetary space.[44]

The "invention" of this device was no mere technical step for the young Tsiolkovsky, but an emotional experience of profound impact, a "peak experience,"[45] associated with intense joy:

> I was so worked up that I couldn't sleep all night—I wandered about the streets of Moscow, pondering the profound implications of my discovery. But by morning I saw my invention had a basic flaw. My disappointment was as strong as my exhilaration had been. That night had a lasting impact on me. Now, thirty years later, I still have dreams in which I fly up to the stars in my machine, and I feel as excited as on that memorable night.[46]

Another translator renders the first phrase, "I was so agitated, nay, shaken...."[47] Yet Tsiolkovsky was not so committed to his "invention" that he would refuse to put it to the test of the physics and mathematics he had taught himself. Another autarchic intellectual, Norman Dean of Washington, D.C., was not so prudent. In 1956 he submitted a patent application for a device very similar to the one Tsiolkovsky imagined around 1873, and three years later received patent 2,886,976 for it. An article by John W. Campbell, Jr., editor of the magazine *Astounding Science Fiction,* described the device in some detail, and presented photographs of a "working" model.[48]

Powered by a conventional electric hand drill, the machine consists of a framework in which two counterrotating weights are pumped up and down. With each pumping motion, there is supposed to be a net thrust upward. For test purposes, the device is placed on a common bathroom scale. It does not fly around the room, because the thrust is less than the total weight of the system, but becomes lighter while sitting on the scale. One photograph shows the machine 16 pounds lighter according to the dial on the scale.

Actually, the device does not work at all. A spring scale, such as the bathroom scale used in the tests, responds nonlinearly to vibrations. If you stand on a bathroom scale and move your fists up and down vigorously, you will appear to lose weight. In fact, you don't. Someone in NASA read Dean's report on the device and concluded that his mathematics was unsound.[49] The Dean machine has not achieved any acceptance by conventional science; it has generated great favorable interest in science fiction circles. Several science fiction fans I interviewed at conventions urged NASA to experiment further with the device. John W. Campbell, Jr., a disappointed, incompletely educated physicist, has complained at length that the only thing holding back spaceflight using the Dean Drive, is the hidebound nature of the scientific community. He has gone so far as to suggest a "two-party system" of science, which would be more tolerant toward new ideas.[50] Campbell agrees that the Dean Drive violates "the laws of Physics," but decries the fact that such laws were ever enacted and advocates their repeal:

> It can't drive in space without drastically rearranging the Law of Conservation of Momentum, and the law of action-and-reaction. And anything that leaks through the Law of Conservation of Momentum automatically challenges the Law of Conservation of Energy. The laws of thermodynamics are based solidly on those; invalidate, or even seriously challenge them, and thermodynamics is a structure without a foundation.[51]

Dean is not in any way a trained professional scientist, not even an autodidact like Tsiolkovsky. Dean was a professional bureaucrat, Tsiolkovsky a teacher. Dean's status of complete outsider to science led him to "invent" a valueless absurdity. Tsiolkovsky's marginal status led him to invent things of little interest to his more conventional colleagues; but they were based on correct principles. Like the outsider pseudoscientists they were, Campbell and Dean conducted a guerrilla war against science. Campbell explains Dean's position:

> ...in his machine, he is a hobbyist at work—an amateur. He's so much of an amateur that, unlike the professional, he could, and did, challenge the fundamental assumptions of physics. Being an amateur, he does not have any appreciable emotional investment in the validity of Newton's Laws; he had no block against challenging them.[52]

On October 19, 1899 Robert H. Goddard, America's most famous spaceflight pioneer, "invented" an interplanetary engine essentially the same as those of Tsiolkovsky and Dean. Goddard recalled this moment, when he was 17 years of age, as decisive for the course of his life:

> I climbed a tall cherry tree at the back of the barn and, armed with a saw and hatchet, started to trim the dead limbs from the tree. It was one of those quiet, colorful afternoons of sheer beauty which we have in October in New England and, as I looked toward the fields to the east, I imagined how wonderful it would be to make some device which had even the *possibility* of ascending to Mars, and how it would look on a small scale if sent up from the meadow at my feet.
>
> It seemed to me then that a weight, whirling around a horizontal shaft and moving more rapidly above than below, could furnish lift by virtue of the greater centrifugal force at the top of the path. In any event, I was a different boy when I descended the ladder. Life now had a purpose for me.[53]

In another recounting of the story, Goddard expressed the idea of the last sentence: "Existence at last seemed very purposive."[54] At last! Goddard's invention was made during a *peak experience,* similar to a religious conversion experience. But, like Tsiolkovsky and unlike Dean, Goddard was not so committed to his new idea that he would not give it a good test. Percy Long, a Harvard freshman friend of Goddard's, tried to convince the boy that the device was impossible but, not being scientifically trained, was unable to marshall decisive arguments.[55] Good Yankee that he was, Goddard attempted practical experimental tests of the idea by making a number of working models—or I should say *unworking* models. He tried every concept that came to his imaginative mind. His father had given him a copy of a popular science book that described Newton's Laws. While Dean had ignored the ideas of conventional physics, young Robert Goddard constructed various devices to test them. Newton's Laws won out, and the centrifugal-drive spaceship lost.[56] Goddard's and Tsiolkovsky's minds were those of uncommitted inventors. Dean and Campbell, for all their protestations of objectivity, were committed against conventional science.

There is no comparison between Goddard's and Tsiolkovsky's family social status, but both had inventors for fathers. Goddard grew up in middle-class, perhaps upper middle-class, comfort. Tsiolkovsky was scorned by his classmates; Goddard was esteemed. Twice elected class president in secondary school, he was the class orator at graduation.[57]

One interesting similarity was that Goddard, like Tsiolkovsky, had become oriented toward intellectural achievement by persistent illness.[58] Throughout much of his working life he struggled against tuberculosis, which finally caused his death. The problem of maintaining self-esteem and forstalling boredom in a sickbed undoubtedly contributed to Goddard's choice of career. He did not suffer radical social isolation, although as his career advanced, he tended to isolate himself from potential colleagues.

Goddard continued his education, received a doctorate in physics, and in time became a professor of physics at Clark University in Massachusetts. Although he recognized the potential of the rocket for space propulsion, he thought that such devices would be far too expensive and it was "hardly worth spending time on them...unless there is certainly no other way."[59] No other way presented itself. On June 9, 1909 he first considered the possibility of liquid-fuel rocket engines.[60]

For a time he supported his research out of his own pocket, but in 1916 he attempted to convince the Smithsonian Institution to support him.[61] As it happened, the Smithsonian had some funds specifically earmarked for atmospheric research.[62] Furthermore, it was in the late stages of a competition for status in the field of aeronautical achievement. The Smithsonian had supported Langley's relatively unsuccessful airplane research and had been beaten to the first true heavier-than-air flight by the independent Wright brothers. For years the Smithsonian had tried to discount the importance of the Wrights' achievement. This time they did not want to lose the prospect of a splended public success.

Goddard wrote a monograph in which he described the potentialities of the high-altitude research rocket, complete with detailed calculations, and submitted it to the Smithsonian.[63] The mathematics in the paper was carefully studied and proved correct.[64] Over the next few years the Smithsonian gave Goddard $11,000 and published this landmark monograph *A Method of Reaching Extreme Altitudes.*[65]

He designed and built the world's first liquid-fuel rocket, which was launched March 16, 1926 at Auburn, Massachusetts. One might be tempted to say that this vehicle was the ancestor of all subsequent liquid-fuel rockets, yet when the Germans were first firing their own liquid-fuel engines in 1930, they did not even know of Goddard's achievements. He did not publish an account of his first 1923 engine firing until 1936, when the V-2 was already taking shape on German drawing boards.[66] Complete reports of his experiments in the 1930s were not published until 1961, the year that Yuri Gagarin orbited the earth.[67]

Goddard received 83 patents for various developments in his rocket work; the executors of his estate culled his notes and found material for 131 more patents.[68] In 1960 the United States government awarded his estate the sum of $1,000,000 "for past infringement and future licences for the future of Goddard's contributions to rocketry."[69] This award not only settled court cases, but established the myth of Goddard's *and America's* priority in space development. Unfortunately, throughout his career, Goddard worked in such extreme self-imposed isolation that his discoveries had little impact.

Reports in the press ridiculed Goddard's rocket experiments and tended to drive him further into hiding.[70] They also had the positive effect of interesting the famous aviator Charles A. Lindbergh in Goddard's work. In the fall of 1929 he visited Goddard, then attempted to interest one of the Du Pont family in sponsoring further research.[71] This effort was not successful, but Lindbergh did interest philanthropist Daniel Guggenheim in the rocket work. Over the next 10 years the Guggenheim Foundation invested $148,000 in Goddard's research,[72] which allowed him to set up a development shop and test base near Roswell, New Mexico. Goddard tended to ignore the work of other men in the field, was remiss in his correspondence with colleagues, refused to share his results, and would not participate in joint projects. He seemed to want to achieve successes, something like the Wrights' workable planes or Edison's perfected working models, then burst upon the world in triumph. Rocket technology is by nature too complex to be perfected in the style of the turn-of-the-century independent inventor. By the time of his death in 1945, Goddard had fallen appreciably behind other groups of American rocket experimenters, and of course the Germans had forged far ahead.

The great man who really mattered in the history of space was Hermann Oberth, born in 1894, in Transylvania. From his earliest teen-age years Oberth was interested in spaceflight and continually returned to Jules Verne's idea of a space cannon that would fire a manned projectile around the moon. As his education progressed, he came to understand better and better the flaws in Verne's plan, to attempt redesigns of it, and eventually to consider alternate means of propulsion, including his own version of the centrifugal drive:

> Then I thought of a wheel with heavy weights sliding to and fro and to on its spokes, the weights being guided by a fixed eccentric ring, so that they would be at a distance from the center on the one side and would come closer to the center on the opposite side. In this way, I thought, a centrifugal force would develop on the side where the weights were at a greater distance from the axis, and lift the wheel.[73]

Oberth's family was economically comfortable, with a summer home and servants. His father was a prominent doctor, and for some time Hermann was expected to follow in this profession. One could not claim, in any simple way, that Oberth suffered blocked opportunity and innovated to achieve status that would otherwise be denied him. Two conceivable modifications in such a theory might render it more plausible. First, Hermann might have resented the pressure to become a duplicate of his father; certainly he did abandon his medical education. He might not have perceived following in his father's footsteps as constituting an opportunity. Second, Hermann turned to spaceflight as a life's interest when he was only 14. At this age opportunities for high status are blocked by time itself. He could not complete medical training for 10 years, and another decade might pass before he could achieve high professional status. By thinking new thoughts, by engaging in a kind of practical status magic, he could attempt to achieve and excel immediately.

Oberth's biographers do not even hint that he was socially isolated or in any way deviant in his personal behavior. Reports of his childhood are, however, filled with incidents indicating an intense desire to excel, for example, in the achievement of difficult feats of swimming. In 1909, at the age of 15, he suffered a debilitating illness, and according to his chief biographer Hartl, his ambition had to be turned from physical achievements toward intellectual excellence.[74] Thus, like Goddard and Tsiolkovsky, his career was shaped by illness. Unlike them, he did not shrink from public activity on behalf of his favorite project. Wernher von Braun, Oberth's chief protege, correctly reports, "Oberth was always active in movements to forward the use of rockets, and encouraged others to work, to lecture, and to write on rocket and space flight."[75]

One can still find grounds to describe Oberth as in some way "deviant." Unlike Goddard, who performed his doctoral research "On the Conduction of Electricity at Contacts of Dissimilar Solids," a safe topic that had nothing to do with spaceflight and which indeed displayed Goddard's training and genius to good advantage,[76] Oberth presented his first monograph on rocket flight as his doctoral dissertation:

> In 1922, I handed my paper to the Heidelberg University as a thesis for my doctor's degree, but it was rejected. I refrained from writing another one, thinking to myself: Never mind, I will prove that I am able to become a greater scientist than some of you, even without the title of a doctor. [In the United States I am often addressed as a doctor. I should like to point out, however, that I am not such and shall never think of becoming one.][77]

Using money partly borrowed from his wife, Oberth published this rejected thesis in 1923 under the title *Die Rakete zu den Planetenräumen* (The Rocket into Planetary Space).[78] In subsequent editions it became the textbook for the German rocket experimenters. Arthur C. Clarke has said that Oberth's book "may one day be classed among the few that have changed the history of mankind."[79] I so class it. Had such a book appeared only a few years later, the German wing of the Spaceflight Movement would have been delayed so much that it would not have been able to exploit the Third Reich, and the chain of events that took Armstrong to the moon in 1969 would have been broken. *Die Rakete zu den Planetenräumen* was distributed to bookstores throughout German-speaking Europe and found its way into the hands of those men who later built the German spaceflight effort.

Like Tsiolkovsky, Oberth may have been gripped by an obsession with weightlessness. As a child he experimented with short periods of weightlessness experienced while jumping from a height into water. He simulated the sensations of zero gravity by floating in water. As an adult, while serving as a military medical orderly, Oberth experimented with "psychological weightlessness" induced by the use of the drug scopolamine.[80]

Oberth's biographers fail to mention that his rocket work was conducted simultaneously with the development of a theosophical system that must be described, delicately, as variant, if not deviant. He invented a number of vitalistic notions, including a doctrine that each cell in the body has its own immortal soul. This led him to a belief in reincarnation. In a sense, when Oberth worked toward an interplanetary future culture, he was preparing a world to dwell in after his death, reincarnated perhaps as a spaceship captain of the distant future! In 1930 Oberth published a pamphlet containing his occult ideas, incorporating the theories he had developed over the previous eight years. A book was ready for publication in 1938, but a series of difficulties prevented publication until 1959.[81] In letters to Max Valier, Oberth urged that publications on rockets and spaceflight should not contain spiritualistic ideas so that the technical content would not be misunderstood or discredited.[82] This was a wise strategy. In effect, Oberth conducted two creative careers, one successful, the other not. He described the intended effect of his 1959 book *Stoff und Leben* as helping people to recognize "that world events are perhaps not as meaningless and purposeless as is commonly considered today."[83] Only as an aside, near the end, does he discuss spaceflight:

> ...and probably Mankind will even build spaceships sometime, make other planets habitable, or even establish habitable stations in space, and having become morally mature in the meantime, will bear life and harmony out into the cosmos.[84]

In his spaceflight writings Oberth has looked forward to a time when humanity will spread across space to inhabit planets in orbit around the most distant stars:

> And what would be the purpose of all this?
> For those who have never known the relentless urge to explore and discover, there is no answer. For those who have felt this urge, the answer is self-evident. For the latter there is no solution but to investigate every possible means of gaining knowledge of the universe.
> This is the goal:
> To make available for life every place where life is possible.
> To make inhabitable all worlds as yet uninhabited, and all life purposeful.[85]

Generation 2: Von Braun and Korolyov

The two most important members of Generation 2 are the German Wernher von Braun and the Russian Sergei Korolyov. Unlike members of Generations 0 and 1, these men were unimaginative and conventional. Neither is described in any historical account as the inventor of anything, but often as the chief designer of one or another rocket vehicle. They were competent engineers and highly successful leaders. If we are to believe the published histories of the Russian program, Korolyov was the single most important member of the Russian branch of the Spaceflight Movement; only Tsander, who died prematurely, stands as a rival. Von Braun was the essential man of the German rocket program, which itself was the essential program.

Both men were extroverted mesomorphic sportsmen; both were active glider enthusiasts. Von Braun was a bona fide member of the Junker upper class. Korolyov was what we might call a member of the intellectual middle class. His biological father was a teacher of literature, his stepfather, an engineer. They were not outsiders nor deviants nor marginal men. They brought the wild ideas of Generation 0 and the abstract calculations of Generation 1 to reality by manipulating the conventional institutions of society and exploiting whatever strains they could find within the conventional power structures. One of the German rocket engineers who came to America with von Braun explained to me that his chief talent was being able to get along with the powerful, to impress them, and to gain their support.

We do have a good deal of trustworthy biographical information about von Braun. His father was a very aloof bureaucrat and statesman; the elder von Braun's autobiography says almost nothing about his sons until they were advanced in their adult careers.[86] His mother, however, was extremely close to Wernher and his two brothers. She was a font of culture, a musician and an amateur astonomer who taught appreciation for the delights of the mind. She provided Wernher with the spaceflight books of Verne and Wells. When the boy was confirmed in the Lutheran church at the age of 14, she presented him with a telescope. The rocket automobile experiments of Max Valier were very much in the news in the mid-1920s, and Wernher imitated the full-scale tests with a toy car propelled by a fireworks rocket. Dangerous explosions resulted.[87] He did not yet know that Valier was merely trying to publicize exactly the rocket theories of Oberth, but he did begin to think about space flight.

Von Braun was not a particularly able student at first. According to one account, he was sent to a private school precisely because his father wanted to channel his energies into more acceptable activities. Wernher "flunked both mathematics and physics,"[88] and was sent away to a boarding school with a reputation for excellence and stiff discipline. In an astronomy magazine Wernher happened to see an article by Oberth describing rocket flights to the moon. Because this topic fitted into his earlier interests in astronomy and rockets, the boy excitedly sent away for a copy of Oberth's great book *Die Rakete zu den Planetenräumen*. He was disappointed to find it a highly technical work filled with unintelligible mathematics. With some encouragement from his teachers, he devoted considerable energy to acquiring the math background necessary to understand Oberth.[89]

Wernher's school habits were transformed and he studied with great zeal. After another transfer to a third pirvate school, he consistently achieved high grades in all subjects and led his class in mathematics and physics.[90] This success undoubtedly increased his commitment to spaceflight.

In June 1927 a group of enthusiasts inspired by Oberth's writings founded the *Verein für Raumschiffahrt* (Society for Space Travel) in Germany. Pendray reports:

> Within a year the society had more than 500 members; by September, 1929, the number had increased to 870. It rose to more than 1,000 soon afterwards; a measure of the European interest at that time in matters interplanetary.[91]

Wernher joined the *VfR* and began to associate with other members. He wrote a letter to Oberth and received a polite reply. Later, Willy Ley, spaceflight journalist and key member of the *VfR*, introduced the young man to Oberth. From this time on Wernher von Braun was one of the central figures in the Spaceflight Movement, rising quickly to leadership, never wavering in his dedication.

Two features of this biography reflect patterns common to a number of other early members of the Movement: conversion to the cause of spaceflight during adolescence, even during times of crisis of self esteem or identity, and recruitment to the Movement through magazine articles or books.

In the Soviet Union more than one rocket organization came into being around 1930. Exactly how "official" or how "amateur" these groups were is not clear, but they certainly received little government support, and the enthusiasm of a few men interested in spaceflight seems to have been the chief force behind them. The best-known space organization was GIRD, the Group for the Study of Reactive Propulsion, which consisted originally of two GIRDs, one in Moscow, the other in Leningrad. The Moscow GIRD, organized in 1931, was headed by Sergei Pavlovich Korolyov. When it merged with the Leningrad GIRD in 1932, Korolyov became formal leader of the new organization, which was shortly transformed into a thoroughly official, rather than primarily amateur, organization. Even Riabchikov admits that official support was slight in the beginning:

> ...the members of GIRD who were working under S.P. Korolev formed a close-knit group and endured some severe hardships. In those trying days, food could be obtained only with ration cards, and very few of these were issued to GIRD.[92]

According to Michael Stoiko, "GIRD members were known to have been refused ration books because they were accused of being occupied with "nonsensical fantasies."[93] Yet by the late-1950s when the first *Sputnik* had been launched and a number of well-financed official rocket development groups were building the ICBMs and space rockets that gave the USSR a temporary lead in space, the situation was quite different:

> Some of the top people in these groups were the young space enthusiasts of the 1920s and 1930s who had dedicated their lives to the development

of rocketry and space travel. They had now become the backbone of the Soviet rocket technical movement.[94]

Our information about the life of Korolyov is thoroughly unreliable and painfully incomplete. We do know that of all major spaceflight pioneers, he was the one most closely associated with aviation. Like von Braun, Korolyov had an interest in gliders. In addition to becoming a pilot and leader of a gliding organization, he designed a highly successful glider in collaboration with the well-known Sergei Illushin.[95] Korolyov was fortunate to have another prominent Russian aircraft designer, Andrei Nikolayevich Tupolev, as his thesis advisor.[96] Tupolev was not entirely free to assist Korolyov throughout his career; this world-famous aviation pioneer was forced to spend the years 1938–1943 in a Soviet prison and even had to design aircraft under those conditions.[97] Vladimirov contends that Korolyov was also imprisoned.[98] In the period 1931–1935 he designed and flew rocket-assisted gliders and turned to rocket-powered cruise missiles, not unlike the later German V-1; he then experimented with rocket interceptor aircraft. By 1947 Korolyov was chief designer of the Experimental Design Bureau, and Soviet accounts report that he led the thrust toward the conquest of space.[99]

All accounts, pro-Soviet and anti-Soviet, agree that Korolyov was the single most important individual and suggest that his influence on Soviet rocket development was decisive. But we have no independent way of determining what really happened, and we depend on romantic Russian popular histories for almost all our information. Korolyov did seem to manipulate the powerful to achieve his own end: spaceflight. Vladimirov praises Korolyov's

> ...capacity for diplomacy. He was very skillful in his approach to the people in power and he knew how to present the course and results of his work in such a way that his bosses felt themselves convinced of the importance of the problems he was dealing with.[100]

Commenting on a number of Korolyov anecdotes that had been appearing in Soviet publications, *New Scientist* said: "Korolyov was prepared to take responsibility, where others might dither or want to pass the buck, and being prepared to do that, he must certainly have pushed the Soviet space programme along."[101]

The Pioneers as a Class

About 100 men played roles of some significance in spaceflight development prior to the creation of military missile programs. In addition to the seven we have discussed, there were a few independent authors and publicists like Valier and Ley. Each of the early rocket clubs had anywhere from six to a dozen really active members. The total membership of these clubs was much greater than 100, but most of the members contributed little more than membership dues and encouragement. The five men we considered in Generations 1 and 2 were more important than the rest by an entire order of magnitude, but the others deserve some consideration as well. In later chapters we discuss some of them as individuals and chart the development of the groups to which they belonged. Here we compare a few of the characteristics of 14 selected pioneers, the seven we have discussed plus seven of the best-known men of the second rank.

I have summarized information taken from a wide range of biographical sources in Tables 2.1, 2.2, and 2.3.[102]

What do these charts show? First, we see that these men were not born in any one area. One of two kinds of spurious effect may be creating this impression: When one pioneer became known in a given area, others might be less likely to take the pioneer role, or be seen as pioneers; and pioneers may be retrospectively "invented" rather than discovered by nations that later develop large space programs; had Italy a space program it might discover it had had pioneers. But, looking at the total historical record, the development of spaceflight was not a phenomenon equally distributed over the globe. Central European nations were of disproportionate significance in the early days. There may be an explanation for this, but some simple notion of industrial or scientific advancement would not explain the Slavic activity, nor the French and British inactivity. Central and Eastern Germany and the Slavic countries were industrially developing areas. Furthermore, they were "encircled" by the Western colonial powers, and perhaps a good deal of the political as well as technological history of Europe in this century can be explained as an attempt by these peoples to gain parity with or superiority over the colonial powers.

Sociologist Nathan Keyfitz has suggested to me that Germany made such great strides in chemistry, before the end of World War II, because it did not possess far-flung colonies or other territories from

TABLE 2.1 SPACEFLIGHT PIONEERS: ORIGINS

Name	*Year Born*	*Birthplace*	*Father's Occupation*	*Family Religion*	*Family Social Class*
Clarke, Arthur C.	1917	England	Simple farmer		Lower
Dornberger, Walter	1895	Western Germany	Pharmacist		Middle+
Ehricke, Krafft A.	1917	Berlin	Dentist	Christian	Middle
Ganswindt, Hermann	1856	East Prussia	Business owner		Upper
Goddard, Robert H.	1882	Massachusetts	Small-factory owner	Episcopalian	Middle+
Kibal'chich, Nikolai	1854	North Central Ukraine	Village priest	Russian Orthodox	Peasant
Korolyov, Sergei	1906	Ukraine	Teacher		Middle
Ley, Willy	1906	Berlin	Wine merchant	Lutheran	Middle
Oberth, Hermann	1894	Transylvania	Physician (senior)	Lutheran	Upper middle
Sänger, Eugen	1905	Bohemia	Merchant		
Tsander, Friderikh	1887	Riga, Latvia	Physician		Middle
Tsiolkovsky, K.E.	1857	West Russia	Forester	Freethinker	Peasant
Valier, Max	1895	Austrian Tyrol	Confectioner		Lower middle
Von Braun, Wernher	1912	Eastern Germany	Cabinet Minister	Lutheran	Upper (Junker)

TABLE 2.2 SPACEFLIGHT PIONEERS: CHILDHOOD AND UPBRINGING

Name	*Early Space Dedication?*	*Science Fiction an Influence?*	*Amateur Astronomer?*	*Recruit-ment*	*Education*
Clarke	Yes (age 13)	Yes	Yes	Mass media	B.Sc.
Dornberger	No	No	No	Army orders	Doctorate
Ehricke	Yes (age 12)	Yes	Yes	Mass media	Interrupted*
Ganswindt	No?	Possibly	No	Self	Dropout
Goddard	Yes (age 17)	Yes	Slightly	Self	Doctorate
Kibal'chich	No	?	No?	?	Irregular
Korolyov	No	?	?	?	Engineering
Ley	Yes (age 20)	No	Yes	Mass media	Interrupted*
Oberth	Yes (age ca. 10)	Yes	Yes	Self	Interrupted*
Sänger	Yes (age 15)	Yes	No	Mass media	Doctorate
Tsander	?	?	Yes	Probably mass media	Engineering
Tsiolkovsky	Yes	Yes	?	Self	Autodidact
Valier	Partly	No	Yes	Mass media	Irregular
Von Braun	Yes (age 14)	Yes	Yes	Mass media	Doctorate

*These men have performed work equivalent to that required for an earned doctorate and have received various honors.

which natural resources could be taken and had to synthesize materials that the colonial powers could simply gather up.[103] Daniel Bell has written:

> At the start of World War I, hardly any of the generals of the Western Allies anticipated a long war, for they assumed that the effective naval blockade of the Central Powers, thus cutting off their supply of Chilean nitrates, would bring Germany to her knees. But under the pressure of isolation, Germany harnessed all her available scientific energy and resources to solving this problem. The result—the development of synthetic ammonia by Bosch and Haber—was a turning point, not only in Germany's capacity for waging war, but also in the connection of science to technology.[104]

In Table 2.1 we see a wide range of fathers' occupations, but one thing stands out. The fathers were *independent.* They worked without close supervision. Most were not manual workers but labored with their minds as well as hands. Three fathers, those of Clarke, Korolyov, and

TABLE 2.3 SPACEFLIGHT PIONEERS: CAREERS

Name	*Occupation*	*Published at Own Expense?*	*Active Publicist?*	*Govern-ment Support?*	*Major Achievements*
Clarke	Popular writer	No	Yes	None	Popularization; British Inter-planetary Society
Dornberger	Military officer, corporation executive	No	No	Much	Military director of German rocket programs
Ehricke	Rocket designer, theorist	No	Yes	Some	Atlas; Space Shuttle
Ganswindt	"Inventor"		No	None	"Prophet" (see text)
Goddard	Teacher; scientist	No	No	Slight	First liquid-fuel rocket
Kibal'chich	Revolutionary	No	No	None	"Prophet" (see text)
Korolyov	Aviation and rocket design; executive	No	No	Much	Sputnik I, first men in orbit, lunar probes, etc.
Ley	Popular writer, teacher	No	Yes	None	Popularization; Verein für Raumschiffahrt
Oberth	Teacher; scientist	Yes	Yes	Slightest	Groundwork for entire Space-flight Movement
Sänger	Aviation designer	Yes	Some	Some	Designed rocket planes
Tsander	Aviation engineer		Yes	None	Organized Russian space organization
Tsiolkovsky	Teacher	Yes	No	Slightest	Laid theoretical groundwork
Valier	Popular writer	No	Yes	None	Publicity; rocket cars
Von Braun	Rocket designer and executive	No	Yes	Much	V-2, Redstone, Wasserfall, Jupiter, Saturn I and V

Valier, died when the boys were small children. Goddard's and Tsiolkovsky's fathers were part-time inventors. The two physicians, the pharmacist, and the dentist had occupations on the margin of science and technology and may have encouraged their sons' own interests in these fields. Ehricke's father and mother were both dentists, and the young boy often used to enjoy experimenting with their exciting equipment.

No tendency for the rocket pioneers to come from religiously marginal families is apparent. Of those about whom we have information, only Tsiolkovsky's was not of the most respectable religious denomination in the geographic area. Many important nuclear physicists in the German scientific community before Nazism were Jewish; I have no evidence that any prominent members of the European Spaceflight Movement were Jewish. Although it is unlikely that no Jews participated and we know of quite a few Jewish science fiction authors, as a group Jews did not have a great role to play in the Spaceflight Movement. When the Nazis came to power, there was no exodus of rocket engineers and spaceflight enthusiasts from Germany.

Table 2.1 shows a wide range of social classes represented among these pioneers. The descriptions are very crude ones, based on the general impressions conveyed by the pioneers' biographies. As a whole, the group appears to be of fairly high social class, including one member of the commercial upper class and one bona fide titled Junker (von Braun). The men certainly have a uniformly educated, middle- to upper-class style. Many of the families were positively oriented toward high culture, and most were financially comfortable enough to afford cultural luxuries such as a piano or opera tickets. With the exception of Clarke and Valier, whose fathers died prematurely, the possibly lower-class men stem from what might be described as industrially underdeveloped parts of the continent where the middle class probably was not large.

In Table 2.2 we see that more than half the men had a strong commitment to spaceflight before reaching the age of 21. One attractive hypothesis is that they joined or created the Movement as a solution for adolescent identity crises. As members of a great ideological movement, they satisfied adolescent longings for transcendent meaning and high status, at least in their own eyes.

A vital part of the cultural milieu surrounding the Spaceflight Movement at its birth was the state of planetary astronomy at the time.

Enough was known and publicized about the solar system for even the common man to know that the planets were "worlds"; enough remained unknown or unpublicized that imagining that the planets might be very much like the earth was possible. In the 1920s educated people could still take seriously Lowell's view that Mars was habitable or even inhabited.[105] Today the planets are known to be inhospitable; in 1925 they beckoned.

Although most of the pioneers were interested in astronomy and owned telescopes in their teens, only Valier became an active and careful amateur astronomer. By the age of 15 he was already interested in such niceties as the resolution of double stars and published articles on astronomy before completing his teens. But the astronomical interests of most spaceflight pioneers were superficial, romantic, and unscientific; the interest of professional astronomers in spaceflight was nonexistent. Not even Lowell is mentioned in the pioneers' biographies. Valier himself never approached astronomy in a truly scientific fashion, and he became a supporter of the *Welteislehre* of Hanns Hörbiger. In this pseudoscientific theory the untutored Hörbiger claimed that space was filled with chunks of ice, seen as the milky way, which continually bombarded the sun, causing sunspots and prominences.[106] For some time Valier traveled Europe lecturing on the merits of this crank idea; he discarded it as his lecture topic only when he was able to replace it with a spaceflight gospel derived from Oberth's work.

Science fiction writer Poul Anderson has proclaimed, "any planet is a *world*...."[107] Superficially, this appears to be a tautology; examined closely, it is either a very profound truth or a grievous error. I suggest that it is an error. Anderson means that every planet is a *Lebenswelt,* a rich environment proper for men to live in, similar but for our absence to the world in which we live. The attitude of astronomers is that every planet is an *object* of study—an object, not an environment seen from the perspective of a potential inhabitant, but a chunk of matter understood from outside as permanently beyond the scope of our personal activity. The moon is not a world; it is a rock.

Astronomers discovered that Earth is a planet. Therefore, in the popular mind, especially that of the 1920s, the other planets are other earths, *other worlds.* Our world, rather, is *on the earth,* encompassing a really small fraction of the planet as a whole, a superficial phenomenon, a very special environment experienced other than as the surface of a sphere of iron. The sensual antiscientific science fiction author C.

S. Lewis continually speaks of his interplanetary characters as being *in* the Earth or *in* Mars, when they stand on the surfaces of these planets. A *Lebenswelt* is essentially not the *surface* of a planet, but rather a biologically active *volume* that encompasses a living person. Rocket pioneers, like many of the general public but unlike most astronomers, have projected a sense of *Lebenswelt* onto their concept of *planet* and always conceived of the purpose of spaceflight as traveling personally to other worlds. I suggest the hypothesis that true activity within the culture of academic and professional astronomy would tend to reduce (rather than increase) the likelihood that an individual would come to support the Spaceflight Movement.

A number of the pioneers did have some experience in aviation. Valier was a World War I air observer. Sänger, Tsander, and Korolyov designed aircraft. Dornberger and Ehricke collaborated on the first designs of the Space Shuttle, an aircraft-like orbital vehicle. Kibal'chich and Tsiolkovsky attempted to sketch various flying machines, and von Braun not only was a glider pilot, but attempted to adapt his rocket engines to aircraft propulsion.

The pioneers who did not come to the idea of spaceflight on their own (self-recruitment) were alerted to the idea and brought into initial contact with other enthusiasts by the mass media: books like Oberth's, newspaper accounts and magazine articles, and even motion pictures. Dornberger was ordered by a superior officer to examine the German rocket clubs for war potential. But none of these men were recruited through already-existing social networks.

What, if any, forces operating in the general society guided the pioneers to form and join the Spaceflight Revolution? One cannot say that European society cried out in the 1920s and 1930s for spaceflight to begin. On the contrary, informed opinion persisted in the belief that spaceflight was impossible,[108] or if possible, purposeless. One could argue that the "Promethean" values of Western Culture demanded spaceflight, operating as social facts quite independent of the expressed wishes and opinions of the individual people which inhabited it. However, the cultural support for spaceflight could have been subcultural, embodied most directly in science fiction literature, popular since the 1860s.

NASA's official historian Eugene Emme has said that science fiction had a positive influence on the development of spaceflight in the early years of the century, but he has suggested that its influence had declined since then.[109] The well-known sicence fiction personality David Kyle told me he had the feeling science fiction had contributed to technological progress until World War II, but had ceased doing so. A survey of biographies of the space pioneers demonstrates that science fiction had at least inspirational value in the early years of the Movement and the early years of the individual pioneers' lives. Emme claims:

> Robert H. Goddard, the American rocket pioneer, was directly influenced by [Jules] Verne. He re-read Verne's novels each year upon the anniversary of the day [October 19, 1899] he had become dedicated to the exploration of space....[110]

This is a nice story, but not entirely accurate. Speaking of the time in his youth when he became dedicated to spaceflight, Goddard himself wrote:

> In January 1898, there appeared daily for several months in the *Boston Post* the story, "Fighters from Mars, or the War of the Worlds, In and Near Boston." This, as well as the story which followed it, "Edison's Conquest of Mars," by Garrett P. Serviss, gripped my imagination tremendously. Wells' wonderful true psychology made the thing very vivid, and possible ways and means of accomplishing the physical marvels set forth kept me busy thinking.[111]

Lehman's thorough biography of Goddard does not describe him as continually rereading the novels of Verne, but rather those of H. G. Wells. He returned to *The War of the Worlds* many times.[112] On May 3, 1932, a third of a century after he first read Wells' classic, Goddard wrote the author a fan letter,[113] saying it "made a deep impression,' and telling Wells of his dedication to spaceflight. Wells made a polite but perfunctory reply.[113]

Tsiolkovsky believed science fiction had influenced his early dedication to space: "I think the first seeds were sown by the imaginative tales of Jules Verne, which stimulated my mind. I was assailed by a sense of longing, and this set me to thinking in a specific way."[114] The novel *Auf zwei Planeten* by Kant scholar Kurd Lasswitz, first published in 1897, was an important stimulus to the German pioneers.[115] Hermann Oberth has written, "...at the age of eleven, I received from my mother as a gift the famous [Lunar] books...by Jules Verne, which I had read at least five or six times, and finally, knew by heart.[116]

These, then, are some of the characteristics of the early pioneers and the forces that may have shaped their careers. The men's biographies make fairly convincing reading, but unfortunately we can no longer get the really thorough information we would need to test some of the hypotheses that might come to mind. Thousands of other men shared many of the same characteristics as did these 14, yet they never joined the Spaceflight Movement. Still, looking back, the state of technology and the sociocultural conditions of the first third of the twentieth century seem to have been conducive to the development of at least a small marginal Spaceflight Movement. Next we examine how that movement could achieve such success in the second third of the century.

Chapter 3 / THE GERMAN SPACE PROGRAM

In the 1920s leading members of the German branch of the Spaceflight Movement made every possible attempt to gain public support. Through a blizzard of articles and books they sought to extract money and encouragement directly from the general population. Rocket societies were set up to channel resources. Corporations were approached and letters were written to military and government agencies. Oberth thought he had the support of a prominent banker, but it did not materialize.[1] He then got caught up in a publicity scheme for the science fiction movie *Frau im Mond,* exhausting himself in a vain effort to build a working model spaceship.[2] All these attempts to build a mass movement or commercial venture failed utterly. Only when Wernher von Braun and a few associates learned how to exploit the German Army through careful deception did the German Space Program get off the ground. In this chapter we examine exactly how it was done.

The major German rocket society, the *Verein für Raumschiffahrt,* was without doubt the largest and most important prewar spaceflight organization. Freeman Dyson has explained:

> The beginning of the space age can be dated rather precisely to June 5, 1927, when nine men meeting in a restaurant in Breslau founded the Verein für Raumschiffahrt. The V.F.R. existed for six years before Hitler put an end to it, and in those six years it carried through the basic engineering development of liquid-fuel rockets, without any help from the government. This was the first romantic age in the history of space flight. The V.F.R. was an organization without any organization. It depended entirely upon the initiative and devotion of individual members....[3]

Dyson is wrong in blaming Hitler for the ultimate collapse of the *VfR*. The Great Depression of the 1930s was the real culprit.[4] The rocket societies stalled because they ran out of funds, but in any case they could not have grown to the necessary strength for real space experiments, even in the best of times. As Dornberger reports:

> One thing was common to all of them...they had no money. No money to do scientific research work, to do exhaustive testing in adequate facilities, to carry on through development. Forced by lack of money, they fumbled around fusing together first primitive, unreliable samples. These dreamers were convinced that private industry, the Press, public organizations, or even their Government would give them millions and millions of dollars only to bring man out in space, where there is no customer to buy man-made products and where no profit could be made. And, since they did not see the brutal facts, they failed.[5]

Interaction Model of the Movement's Success

Blocked in their attempt to secure civilian and public support for spaceflight, the Movement executed a "military detour."[6] Leaders of the various branches of the Movement were able to exploit governments on many occasions. The majority of cases fit a simple model:

There are three *actors* in the model:

1. The *spaceman*. A leader of one of the branches of the Spaceflight Movement, he wants to bring spaceflight about and is prepared to use almost any means to accomplish this goal.
2. The *patron*. A leading member of the larger society, he has considerable resources at his disposal and is relatively free to spend these resources as he individually decides.
3. The *opponent*. A peer of the *patron*, he is locked in fierce competition with him.

These actors may be organizations rather than persons, but usually are persons, or persons who lead organizations.

There are several distinct steps to the process:

1. The *opponent* gains an advantage over the *patron* in their competition.

2. The *patron* perceives this and seeks a countermeasure.

3. The *patron* comes to the opinion that conventional means may not redress the imbalance.

4. The *spaceman* comes to the *patron,* playing the role of "technical expert" and attempts to sell his own favorite project as though it were a solution to the *patron's* problem.

5. The *patron* invests in the project, typically without an independent assessment of its value.

6. The project does further the aims of the *spaceman;* it may or may not further the aims of the *patron.*

7. If the project does give the *patron* an advantage over the *opponent,* the *opponent* is liable to become the *patron* of another *spaceman.*

8. Given a differentiated social world, where powers are forever locked in mortal competition with other powers, in the context of the technology of the mid-twentieth century; and because many modern political leaders have been technically ignorant and not responsible to their publics; the conditions required for this model probably obtain frequently.

9. However, space technology has been in implicit competition with other technologies for patron support. After some point in the general development of modern technology, long-range rocketry would either have lost its plausibility as a solution to political problems or would have been gradually achieved in such an efficient manner that the necessary misdirection to produce space vehicles under the cover of weapons would no longer have been possible.

10. Thus, the model requires: strategic behavior by the elite of the movement, operating in a conducive political and technical environment, racing with an implicit time limit after which the necessary conditions would have passed away.

To illustrate this model and demonstrate that it fits the historical facts, we consider the development of the German Space Program in terms of 15 cases—more-or-less distinct steps on the way to the moon. The first two are somewhat primitive in that not all the conditions of the model are met; thus, they were not particularly successful. The next dozen cases fit the model well. The final case does not fit the model closely, introduces a new pattern of space development, and therefore marks the end of the Spaceflight Revolution.

CASE 1

PURPOSE:	Initial investment in rocket motors for space applications.
SPACEMAN:	Max Valier
PATRON:	Fritz von Opel (automobile manufacturer)
OPPONENT:	Other auto makers
PATRON'S INFERIORITY:	Not necessarily inferior—simple competition with other manufacturers.

Valier sold Opel on the idea of producing a few test racing cars propelled by rockets. The motors were almost all simple solid-fuel rockets, and there is some question whether the tests advanced or retarded the cause of spaceflight. Valier hoped to generate such great interest that industrialists would make massive investments in rocket development. Public interest was greatly stimulated, but massive financial support never came. When Valier was killed in a poorly controlled engine test sponsored by a commercial enterprise, further support by industry was rendered even more unlikely.[7]

This case is defective in two ways: (1) Opel was probably not in a death struggle at the time, so his commitment was superficial. (2) Rockets were such an inappropriate means of propulsion for cars that further development was implausible.

CASE 2

PURPOSE:	Continued rocket development after the collapse of the *VfR*.
SPACEMAN:	Rudolf Nebel
PATRON:	City of Magdeburg
OPPONENT:	Other German cities and the reputation of earlier burgomaster of Magdeburg, Otto von Guericke.
PATRON'S INFERIORITY:	Not necessarily inferior—general status competition of German cities

In various attempts to salvage the *VfR*, its most active leader, Rudolf Nebel, looked for financial support wherever he could hope to find it.

Current in Germany at the time were a number of crank pseudoscientific theories, including a doctrine known as the *Hohlweltlehre* which held that the true shape of the earth was a hollow sphere and that we lived inside this sphere, rather than on the outside surface of a solid sphere.[8] Our heads, not our feet, point to the center. City Engineer Mengering of Magdeburg was a devotee of this occult dogma and conceived the plan of building a large rocket that would be fired 8000 miles straight up and, to prove the correctness of the *Hohlweltlehre,* would smash again into the earth at the antipodes, without ever curving from a straight course!

Magdeburg had long before supported the famous scientific experiment that publicly demonstrated air pressure when two hemispheres placed rim to rim and evacuated could not be pulled apart by teams of horses. The prestige of Magdeburg over other German cities would be boosted by a new stunt of equal magnitude and public interest.

Nebel negotiated a deal in which some thousands of marks would be invested in building such a rocket. Later designs called for a human passenger. Presumably, when the ship reached a high altitude, he could look out the window and directly *see* the true shape of the earth. If the horizon went up, the *Hohlweltlehre* was confirmed; if the horizon went down, it was disconfirmed. Little came of the project, but as Herbert Schaefer, one of Nebel's associates, reports, at least the *VfR* had a chance to work in a condition better than the accustomed poverty.[9]

CASE 3

PURPOSE:	Early development of liquid fuel engines
SPACEMAN:	Wernher von Braun
PATRON:	The German Army
OPPONENT:	The Western Powers
PATRON'S INFERIORITY:	Germany prohibited from building heavy artillery by the Treaty of Versailles

This case is important, complex, and in some ways primitive. There is an intricate reciprocity between the actions of the *spaceman* and the *patron.* Only after this case does the German Movement understand the way it must go. Consequently, we must examine it in detail.

After defeat in the Great War, the German military renaissance began long before the rise to power of the Nazis, and the rocket program of the German Army was nearly a decade old before the Nazis exerted a major influence on it. Under the vigorous leadership of men like General Hans von Seeckt, the rebuilding of German armed forces

> involved not merely the concealment of weapons and equipment, the training of illegal levees or the surreptitious maintenance of forbidden institutions like the General Staff, or branches, like tanks and aviation. It was an almost fantastic effort to preserve, at least in its basic foundations, the entire huge organizational and institutional machinery indispensible for the effective establishment and functioning of a modern mass army.[10]

Opinions differ on the extent to which the German airline Lufthansa was developed as a prototype Luftwaffe rather than for its own sake and for the benefit of the civilian economy.[11] Much secret work was going into the preparation of a strong and advanced air force from the early 1920s onward. Secret agreements were concluded with the Soviet Union which permitted Germany to set up test and training bases on Russian territory, including a gas warfare school, a tank school, and an air base and training school for military pilots at Wivupal/Lipeszk, 400 kilometers southeast of Moscow.[12] Leading men in the German aviation industry were aware of, and frequently participated directly in, these clandestine efforts.[13]

The prohibition of heavy artillery by the Treaty of Versailles could not be directly violated by the Germans in a secret manner, because a major industrial effort would have been necessary. This restraining document, however, made no mention of rockets, so legally the Germans could develop and produce war rockets in any size and number. Military rockets had played a small but continuing part in warfare since the year 1232.[14] European armies had contained some rocket artillery units since the beginning of the nineteenth century, although the only nation to use them operationally in World War I was France.[15]

To investigate the possibility of developing further the rocket's military potential as a way of circumventing the restrictions against heavy artillery in the Treaty of Versailles occurred quite naturally to some German military officers. This increased interest in rockets was caused by the blocking of conventional means for achieving conventional ends and arose in a situation of extreme German inferiority compared to their opponents. The normal progress of unfettered technological development did not lead to a major research and development investment in the rocket; rather, an abnormal situation encouraged the revolutionary technological advances that followed.

Professor Doctor Karl Becker, Chief of the Ballistics and Munitions Branch of the German Army, chose the brilliant young officer and engineer Walter R. Dornberger to head a small project to evaluate and further develop German rocketry. Dornberger, who was later to be one of the decisive figures in the development of spaceflight and most active leaders in the Movement, had previously exhibited no interest in space nor any faith in the future of rocketry. Dornberger says Becker gave him the following instructions:

> You have to make of solid rockets a kind of weapon system which will fire an avalanche of missiles over a distance of 5 to 6 miles, so as to get an area effect out of it. Next, you have to develop a liquid rocket which can carry more payload than any shell we have presently in our artillery, over a distance which is farther than the maximum range of a gun. Secrecy of the development is paramount.[16]

Interestingly, Becker believed liquid-fuel rockets might have more potential than solid-fuel rockets. We do not know how he came to this opinion, nor what other individuals may have participated in shaping the decision to explore the possibility, but clearly the calculations, experiments, and propaganda of the German spaceflight pioneers, such as Oberth, Valier, and the *VfR,* provided the only background for such an orientation. As we explain in the following chapter, liquid-fuel rockets actually have less military potential than solid-fuel rockets, even though their potential for space vehicles is much greater. By stressing liquid-fuel engines, the elite of the Spaceflight Movement molded the consciousness of all those who contemplated the future of rocketry and achieved a technological definition of the situation that, although objectively incorrect, was favorable to their cause. Becker would have been better advised never to invest in the development of expensive liquid-fuel rockets but to put more emphasis on chemical research on new solid fuels. Whatever the intermediary social interactions, Becker could not have turned to liquid-fuel engines had not the Spaceflight Movement prepared the way.

At about this point in time, around 1930, Rudolf Nebel came to Becker with a written proposal for military development of long-range rockets. Nebel initiated contact between the Army and the *VfR,* although he was not directly responsible for Army interest in the idea. Becker was not impressed by Nebel, whom he saw as an amateur visionary.[17]

In 1930 Dornberger was made assistant to one of Becker's subordinates, Captain von Horstig, who had recommended Dornberger for the project.[18] At first Becker attempted merely to subsidize the more promising of the competing independent rocket researchers, but it soon became clear that progress was not resulting and the Army would have to bring the best men together in an in-house project.[19] Dornberger notes that the Army would have preferred German industry to take over this task, but no company wanted to get involved,[20] particularly after Valier was killed in an explosion while performing publicized tests in collaboration with the Heyland company.[21]

Dornberger surveyed German rocketry and turned up five independent liquid-fuel engine projects. He was not impressed by the experiments of Johannes Winkler with "some small rockets which exploded every time he ignited them."[22] He had some respect for the work of Friedrich Wilhelm Sander, a professional solid-fuel rocket manufacturer of great ability, and was under the impression that Sander's experiments with liquid-fuel engines had achieved some success. Although Sander contributed to the further development of military solid-fuel rocketry, he never participated in the Army's liquid-fuel program. One possibility is that false newspaper publicity had claimed tremendous success for Sander engines long before they had actually worked, thus confusing Sander's reputation.[23] Another completely independent worker, Albert Püllenberg, was not achieving any success and lacked the necessary training.[24] The fourth researcher, Valier, was of course killed just when Army interest was becoming strong.

This left the fifth and largest group, the *Verein für Raumschiffahrt.* Dornberger, von Horstig, and Becker visited the *VfR's* modest headquarters, the Raketenflugplatz Berlin, at the end of 1931 and again at the beginning of 1932. The activities there did not seem sufficiently scientific to the military men. Dornberger complained that the *VfR* was obsessed with visionary fantasies. Much of their work concerned aspects of spaceflight that had no application whatever to current military needs. Members of the *Verein* had their heads not in the clouds but far beyond them.[25]

VfR leader Nebel made an energetic effort to convince the trio of officers to invest in the *Verein,* met with them several times, and attempted to produce an outline of the work that could be done. He failed to impress them. They found him as unconvincing and fanciful as the other spaceflight enthusiasts with whom they had been dealing. Nebel complained to Wernher von Braun, his teen-aged associate, that the Army men lacked imagination and couldn't even realize that the *VfR* was struggling to achieve a technical revolution on a financial shoestring.[26]

Von Braun decided to interject himself into the stalled negotiations and, for the first of several times, played an historically decisive role in furthering the cause of spaceflight. He went directly to Becker and countered charges that the *VfR* conducted its work in a circus atmosphere with the fact that money had to be raised in some manner and argued that any deficiencies of the group were caused entirely by its impoverished state. Dornberger reports the impression the military men formed of von Braun:

> He was the only one of the group not filled with baseless enthusiasm. Displaying great theoretical knowledge, he would realistically tell us what he felt the trouble was and how to correct it. "It won't be easy," he told me. Because of this lack of boasting, this reasonable approach, von Braun was the first man I hired for our own group when it was formed.[27]

Dornberger and his fellow officers were very much aware that von Braun was not just any star-struck youngster. He was the son of Baron Magnus Freiherr von Braun, Minister of Agriculture in the German Cabinet. Willy Ley has remarked that von Braun approximated the Aryan Nordic racial type favored by the Nazis. Ley also commented that the young man's "manners were as perfect as rigid upbringing could make them." Wernher "spoke a rather good French" and could play Beethoven's Moonlight Sonata on the piano by memory.[28]

Under Dornberger's direction, a scientifically designed test stand for liquid-propellant rockets was built at the Army's Kummersdorf artillery range near Berlin. Von Braun, just 20 years old, worked at first with a single assistant developing a successful small engine, getting credit toward his Ph.D., while fulfilling Dornberger's assignments.[29] The first successful engine test was made in January 1933. We do not follow in detail the technical developments from this initial success, through the many setbacks, designs, and test-vehicles, until the first V-2 flight nearly 10 years later, but concentrate on the social aspects of rocket development within the German Army.

From the time von Braun was hired in November 1932 until work at Kummersdorf was completed in 1937, the staff grew to a total of 80 men. Initially, funds were invested in accordance with Becker's modest development plan. As costs increased, von Braun and Dornberger became adept at various strategies for justifying expenses. For example, project expenditures were supposedly limited to the actual costs of building and testing the rockets; there were no appropriations for overhead. So Dornberger and von Braun resorted to pranks and subterfuges to get what they wanted. In purchase orders they described various bits of office equipment in such contorted bureaucratic language that they sounded like scientific instruments.[30] The line between such bending of the truth and outright deception is difficult to determine. At some point they crossed that line.

Two important social developments were occurring at this time: (1) Men like Dornberger and the various engineers being brought into the growing rocket project were being converted to the Spaceflight Ideology and made into loyal members of the Movement. (2) The elite of the Movement were learning strategies of subterfuge, a praxis of deceit, and the exploitation of military and government patrons, which would continue throughout the development of space technology.

The successes of the team's A-2 liquid-fuel rockets, two of which flew at the end of 1934, provided proof that such missiles could be built, and Dornberger and von Braun engaged in a vigorous program to convince Army leaders of the value of this work. They conducted engine tests as theatrical demonstrations for visiting officials, impressing the military men with the overpowering noise, the smoke, the flame, and well considered lectures.[31] The annual budget of the project remained however less than 80,000 marks.

CASE 4

PURPOSE:	Continued development of liquid-fuel rocket engines
SPACEMAN:	Wernher von Braun
PATRON:	Ernst Heinkel (aircraft manufacturer)
OPPONENT:	Messerschmitt (aircraft manufacturer)
PATRON'S INFERIORITY:	Messerschmitt gaining official support in Luftwaffe; Messerschmitt building very fast fighter planes (e.g., Me-109); rocket engines adapted to interceptor aircraft in unsuccessful attempt to surpass Messerschmitt

Undoubtedly recalling Max Valier's plan to solicit support for rocket development by adapting rocket propulsion to high-speed aircraft, von Braun decided to seek an alliance with an aviation industrialist and in November 1935 visited Ernst Heinkel, founder and director of the Heinkel aircraft corporation. Von Braun had no difficulty in convincing Heinkel to provide him with an aircraft in which to test one of his engines.[32] Heinkel himself had been present at one of Valier's rocket automobile demonstrations and had been greatly impressed.[33] At this time Heinkel was competing with Messerschmitt in producing fast fighter planes and record-breaking test aircraft. Rocket propulsion seemed to Heinkel a likely way of leapfrogging the competition's newest propellor aircraft. Later Heinkel supported the work of another independent young man, von Ohain, who was working on the development of the air-breathing jet engine. Of course jet aircraft are far more practical than rocket aircraft for all general uses, and while von Ohain's work led to the first operational jet planes, von Braun's experiments led merely to the demolition of several Heinkel aircraft through explosions and a few demonstration flights.[34]

CASE 5

PURPOSE: Continued engine development and construction of a new and larger test base

PATRON: German Army

OPPONENT: Luftwaffe

PATRON'S INFERIORITY: Luftwaffe gaining favor with Nazi regime; Army losing long-range bombardment role to Luftwaffe planes and wanting to be able to equal its feats; in turn, Luftwaffe leaders looking for new advantages, and Army leaders concerned to top them

The brief collaboration with Heinkel was useful to the Movement in promoting the idea of rocket-powered aircraft. Luftwaffe Major Wolfram von Richthofen visited Kummersdorf to get an impression of the rocket's potential in aviation. Von Richthofen, first cousin of the famous Red Baron air ace of World War I, was in charge of aircraft development in the German Air Force. He would shortly command the infamous Condor Legion in the Spanish Civil War.[35] He gave von Braun an order for initial work on a rocket fighter plane and was impressed by the quick progress made. Von Braun convinced the major that even faster progress could be made if his team had better support, and von Richthofen pledged a huge 5,000,000 marks. Von Braun has commented.

> His offer constituted an unprecedented breach of military etiquette as between branches of the Wehrmacht. My immediate superior, Colonel von Horstig, solemnly led me into the office of General Becker, who had become Chief of Army Ordnance. The General was wrathfully indignant at the impertinence of the Junior Service.
>
> "Just like that upstart Luftwaffe," he growled, "no sooner do we come up with a promising development than they try to pinch it! But they'll find that they're the junior partners in the rocket business."[36]

Becker vowed to top the Luftwaffe's 5,000,000 marks by investing 6,000,000 of the Army's money. With early design versions of the V-2 on the drawing boards, he thought that the large ballistic rocket might give the Luftwaffe's bombers tough competition. The challenge could not be ignored.

In December 1936 Wernher von Braun spent the Christmas holidays with relatives in the north of Germany at Anklam on the Baltic. He was reminded of a forested area where his father had enjoyed duck hunting and came to the opinion that the northern end of a nearby island, Usedom, would be an excellent site for a rocket test station. Dornberger visited the area a few days later and agreed that his was the ideal location to continue rocket development.[37] Over the next several years the German Army invested 300,000,000 Reichsmarks (about $120,000,000) to build, equip and staff this now-famous test center, called Peenemünde.[38]

CASE 6

PURPOSE: Recruitment of scientific and engineering talent to the Peenemünde projects

SPACEMAN: Wernher von Braun

PATRON (recruit): Numbers of German technical intellectuals convinced to contribute their talent

OPPONENT (threat): The German military and its hunger to consume these same men as engineers or common soldiers

PATRON'S INFERIORITY: In the context of a difficult war, with official and public pressure for all able men to fight for the Third Reich, talented men with few means to protect their lives and honors

Despite the secrecy surrounding the Peenemünde projects, von Braun and his associates were able to engage in a variety of public relations and recruiting activities. Over the years many Nazi leaders and military officers including Hitler himself,[39] were brought to Peenemünde to view dramatic displays. From September 28 through September 30, 1939, Peenemünde was host to a secret convention of technical experts from a variety of fields. This important meeting came to be called the *Tag der Weisheit,* or to use von Braun's own translation, Wisdom Day. This event, of course, took place only four weeks after the German invasion of Poland. Thirty-six "professors of engineering, physics, and chemistry," convened to discuss the challenge of the long-range rocket and its subsystems. A number of the participants—including Professor Schuler of Göttingen, a gyro expert;[40] Professor Wolmann of the University of Dresden; and Professors Stasblein and Fassbender of the University of Berlin, experts in signal engineering (telemetry)—were working in fields closely related to missile guidance. According to interviews conducted with some of these men by the U.S. Army in 1945: "During this meeting these scientists were told about the large scale developments in rocket-propelled missiles and were given specific assignments."[41] Von Braun explains the professors' enthusiasm:

> Since the universities were also suffering from [military] conscription the professors were all the more eager to participate in a novel scientific effort which might also place academics in better rapport with the government.[42]

The experts went to their classrooms and laboratories to work on their specific assignments, and later "there were many symposia and mutual visits."[43] Mathematician A. Walther of the *Darmstadt Technische Hochschule* is a good example of a participant protected from the harshest impact of the war by his Peenemünde connections. Walther was actually drafted into the German army, before being "rescued" because of his importance to the V-2 project. He spent just one week in military service before returning to the calculation of trajectories. "Darmstadt was so busy as a result [of the V-2 project] that it could practically guarantee its people freedom from war service."[44]

The military priority rating of Peenemünde's work went up and down throughout the early years of the war, but von Braun could often seek and hire many engineers and technicians. He wrote: "to my intense satisfaction I was able to send out a call for those who had learned to love rocketry at the old Raketenflugplatz,"[45] old members of the *Verein für Raumschiffahrt*. Naturally, these men were already dedicated members of the Movement.

Von Braun travelled frequently to various technical centers throughout Germany, coming to know talented young men and either inviting them immediately to Peenemüde or keeping a record of them for future contact. From a variety of sources, I have been able to find the birth dates of 26 of the 118 Peenemünde men who came to the United States in 1945 and 1946, and the mean is 1912, the year of Wernher von Braun's birth. The standard deviation, 3.25 years, indicates that these men were indeed of one generation. In 1942, when the first V-2 was successfully launched, they averaged just 30 years old. Although these 26 are not a random sample of the whole, they represent an interesting sample, because they are the most important and gifted by von Braun's own estimation.

Sometimes the enthusiasm of the new recruit bordered on religious fanaticism. Dornberger writes with amusement of the day in 1939 that Dr. Steinhoff, then 31, appeared at the base. He had "literally run in." Von Braun had met Steinhoff at the sailplane institute in Darmstadt and invited the young man to visit the rocket base. Steinhoff, contrary to security regulations, was allowed to see an engine test and was so impressed he cried to Dornberger: "Lieutenant Colonel, Sir, take me! You have me body and soul. I want to stay!"[46]

The social transaction of finding these men, selected for high enthusiasm, and convincing them to come to Peenemünde built a tight-knit team of high commitment that was interpreted by members themselves as commitment to spaceflight.

Dieter Huzel is a good example of a member who came to Peenemünde after a long time in military service. Son of an engineer-executive at the Krupp plant in Essen, Huzel himself became a project electrical engineer in the Siemens company. As a youngster he had attended one of Max Valier's spaceflight lectures, but he was not recruited into the Movement until some years later when he was brought to Peenemünde through his friendship with Hartmut

Kuechen, already employed at the rocket base. Drafted by the Army in 1942, Huzel resented the fact that he was taken away from his proper employment to a job where his talent would be wasted: "I was a *Landser,* an ordinary foot soldier, and my real capabilities, along with those of thousands of other good technical people drafted in a similar manner, were lost to the now-desperate German war effort."[47] Huzel had served on the dangerous East Front.

German engineering talent assigned to such perilous and unproductive service was discovered through the most diverse means by von Braun, and a single special maneuver was used to ease bringing the men into the project without appearing to deprive the Army of them. A great percentage of the technical manpower at Peenemünde was officially assigned to a military unit, *Versuchs-Kommando Nord,* Test Detachment North, so the military could imagine they were still serving its purposes. In all practical respects, except pay and mobility, the men had been removed from military service to become civilian engineers. Field-Marshall Walther von Brauchitsch, one of the very few high officers consistently enthusiastic about the V-2 project, assigned a reported 3500 to 4000 such men to Peenemünde: "...nominally to be trained, they were actually used to expedite development."[48] The men could thank the cause of space rocketry for rescuing them from the meatgrinder of the front. Commitment to the value of the German Space Program may have been increased by a potential sense of guilt in being spared from conditions and dangers that so many other Germans were forced to face.[49]

The importance of the Peenemünde men has been shown both by later developments and the attitude taken toward them by Germany's enemies during the war. On August 17, 1943 the British sent 600 four-engine bombers against Peenemünde. Three million pounds of high explosive, the payload equivalent of 1500 V-2s, were dropped on the rocket base, along with many incendiary bombs. Forty-seven British bombers were shot down, and 735 Germans were killed in the raid. Historian Clarence Lasby explains:

> Target priorities boldly emphasized the importance attached to scientists; heading the list were the housing estates. It was hoped that a ten-minute saturation of the residential area would kill the scientists and thereby cripple the project.[50]

Although there was considerable damage to the rocket center, most of the key personnel escaped injury, and it was possible for most development teams to be back at work within a few days.[51] The British attack had been planned and urged by Duncan Sandys. When the *Apollo 11* expedition successfully reached the moon in 1969—boosted to their target by the Saturn V rocket, a direct descendant of the V-2 mainly designed by von Braun's team in America—Sandys cabled his congratulations to von Braun; "Warmest congratulations on your great contribution to this historic achievement. I am thankful that your illustrious career was not cut short by the bombing raid in Peenemünde twenty-six years ago."[52]

CASE 7

PURPOSE:	Final development of the A-4 (V-2) rocket as a demonstration vehicle for the value of long-range rockets.
SPACEMEN:	Von Braun and Dornberger
PATRON:	Adolf Hitler
OPPONENT:	England
PATRON'S INFERIORITY:	Germany no longer able to blitz England with conventional aircraft; the Allies gaining command over German airspace

In its earliest design and development stages, the Peenemünde team's V-2 rocket was financed by the German Army acting independently. Its final development stage and the massive demonstration of its capabilities in combat were financed more directly by the Nazi government. Wernher von Braun has said of the beginning of the project:

> There is not a shred of truth in any statement that the A-4 [V-2] was originally conceived as a weapon with which to devastate London. In actual fact, Dornberger found that the higher Army authorities refused to grant necessary research and development funds after the failure of [the predecessor] A-3 in the summer of 1937 unless there were quite definite prospects of a useful military weapon in sight.[53]

The V-2 was designed to carry exactly one ton of explosives over a distance of more than 200 kilometers, figures picked in great measure because they sounded impressive, particularly in comparison with other rockets or conventional artillery. As both an engineering development step and good propaganda, von Braun's team produced the successful A-5 test vehicle, a small liquid-fuel rocket, before returning to complete the V-2, then still called the A-4. As the military fortunes of the Third Reich rose, the production priorities for the long-range rocket fell. There is a story, possibly fabricated by Albert Speer, to the effect that Hitler himself dreamed that no V-2 would ever reach England.[54] This magical intuition was used as an excuse to withhold high priority from the project. By 1943 von Braun and his team had achieved a few successful test flights and decided to launch a new and aggressive sales effort.

On July 7, 1943 Dornberger and von Braun were able to secure a private audience with Hitler himself. They put on a strong hard-sell lecture, complete with a thrilling motion picture of V-2 launches, stressing the immense power and invulnerability of this rocket. Dornberger had been disappointed when Hitler had earlier witnessed such a launch without being impressed, but now the military situation had worsened, and Hitler was casting about for magical solutions to his problems. This time, after the sales pitch, Hitler excitedly demanded superrockets that would devastate his enemies with "total annihilation." Dornberger reports:

> Hitler stepped quickly up to me and shook my hand. I took in his almost whispered words:
>
> "I thank you. Why couldn't I have believed in your success? If we had only had these rockets already in 1939, then it wouldn't have come to war...."
>
> His gaze seemed to wander off into space. He didn't see me any longer; only his mouth spoke on..."now and for all the future, Europe and the world are too small for a war. With such weapons, a war would be unbearable for humanity."[55]

Bernard Barber has noted the irrationality of the Nazi management of science and technology caused by "their espousal of the cultural value of emotional irrationality."[56] I am arguing here that although *rationality* furthers normal technological progress, *irrationality* is an important precondition for many kinds of revolutionary technological advance. Barber says further:

> Hitler, it has been reported, often counter-manded the advice of his assistants, advice based on rational investigation and planning, in favor of "hunches." Sometimes these half-irrational insights led to success; some-

> times to failure. For instance, because of his irrational desire for miracle weapons, Hitler was susceptible to wild and quackish notions about scientific possibility.[57]

New funds poured into the V-2 project. Literally hundreds of test rockets were fired, and there was even money left over to support a variety of projects to advance still further the cause of spaceflight. After the war General Leslie Simon studied the practice of research by German engineers and scientists who secured contracts from the military and observed that it was quite common for them to succeed in exploiting their "masters."

> On being pressed, the Germans [scientists] admitted that they were taking advantage of military ignorance and of a benevolent government to pursue the kind of research they wished rather than pursuing the kind of research which could be expected to be remunerative in the war effort. It was a case of insufficient scientific competence within the military framework to accomplish properly the task of supervising and judging the scientific work done in the military interest.[53]

On November 11, 1943, in a speech at the Kassel City Hall, Goebbels took notice of the wounds British air power was inflicting on Germany, and promised:

> It will not last too much longer (audience applause); it will not last too much longer until our defense has again become master of the air war, at least has taken it under our control again. And the day will come, in not too long a time, when we will be able to retaliate against England for that which has been done to us over the past months, and, to be sure, in a style (shouts of "bravo!"—powerful applause), and to be sure in a style of which it is plainly impossible for them to conceive today.[59]

Goebbels said Germany would retaliate, *vergelten* in German, and had in mind the Vergeltungswaffen, the V-weapons then under development. Indeed, the person who gave the name V-1 to the Fieseler-103 robot bomb and the name V-2 to von Braun's A-4 rocket was Goebbels.[60] Half believing his own propaganda, he broadcast the myth of wonder-weapons which would turn the tide of the war. The words he spoke before a crowd in the Berlin Sportspalast became a slogan: "*Eines Tages kommt die Stunde der Vergeltung!*" (One day will come the hour of retaliation!)[61] On July 26, 1944, after the first use of the V-1 against Britain but before the beginning of the V-2 attack, Goebbels explained his views on the technological situation to the German people by radio, arguing that the decisive fact about the V-1 was that "it moves in an entirely new context" and "presents the enemy with *absolutely* new facts." He then promised:

> The case will be similar with other novel weapons which we will shortly bring into action in the most varied spheres. We have not merely *overtaken* the lead that the enemy formerly held in this or that sector of war technology, but *surpassed* it.[62]

The German wonder-weapons, including the V-2, did not overcome the great disadvantages of the military and economic situation of the Reich, and developing the long-range rocket into a really worthy weapon remained for the United States and Soviet Union. In the following chapter we examine in some detail the real capabilities and technical implication of the V-2 and its relatives. While Goebbels was able to believe it was an effective weapon, the Peenemünde team did not in fact produce a viable military device; they were able to convince the Nazi government otherwise only through the exploitation of the government's intense needs and weak technological understanding.

This case might be described as rather a collection of many individual cases fitting the same model. In Chapter 4 we discuss the *Wasserfall* missile, sold to the Nazis as an effective air-defense rocket. This device could certainly be described as a separate case, as could the A4b design—a winged V-2—and many component advances and drawing-board or wind-tunnel studies. There is no need to divide this period of German technical history because the material is always the same: Again and again the elite of the German Space Program misled military and political leaders into investing in expensive development programs.

Von Braun's team was not the only group in Germany working on rocket propulsion. Some others had no desire to further the cause of spaceflight and worked on simple solid-fuel bombardment rockets or other modest devices. The liquid-fuel rocket engines developed by a team at BMW for the Messerschmitt-163 rocket interceptor also did not bend its designs in the ideal direction for spaceflight.[63] There was at least one "back-up" spaceflight team: Eugen Sänger's.

CASE 8

PURPOSE Development of a manned space plane
SPACEMAN: Eugen Sänger

PATRON: Luftwaffe

OPPONENT: German Army (von Braun) and to some degree the Allies

PATRON'S INFERIORITY: Army leading in rocket development; Allies leading in long-range bombing; Germany unable to touch the U.S. enemy at all with its moderate-range aircraft

Of all the various secret projects under development in Germany, only Sänger's *antipodal bomber* was in competition with von Braun's attempts to further space travel. Both because of low funding and a too advanced basic design concept, it would not have acted as an adequate back-up to von Braun's V-2, had the Peenemünde project been cancelled before completion. However, it demonstrates the same principles of exploitation and is worth considering.

Sänger was an Austrian aviation engineer who was converted to the Spaceflight Movement between the ages of 16 and 18 by the science fiction of Lasswitz and the theoretical works of Oberth. He had intended to submit a work on rocket technology as his doctoral dissertation, but was convinced by one of his professors to choose a more conventional topic in aeronautics. He received his degree in 1929. Like Oberth, Sänger was forced to borrow money from his wife and others to pay to have his first book *Raketenflugtechnik* published by the same house that had put out Oberth's, Valier's and Hohmann's spaceflight books.[64]

Sänger's approach to spaceflight was aeronautical rather than ballistic; he saw spacecraft as a natural step in the further development of high-speed aircraft, rather than as a completely fresh technological development. Throughout his career he labored to perfect designs for a suborbital, winged, piloted rocket spaceplane and remained faithful to one set of design concepts. In October 1933 he submitted a plan for a rocket-powered high-speed aircraft to the Austrian Ministry of Defense. On February 3, 1934 this institution sent him a scornful reply convinced that liquid-fuel rockets could never be adapted for aircraft propulsion. Working almost on his own and still laboring to pay off the debts incurred in printing his book, Sänger developed his own rocket engines, testing the first models successfully in mid-1934.

In 1936, two years before the *Anschluss* in which Germany annexed Austria, Sänger was invited by the German Air Ministry to perform research at the German Experimental Center for Aeronautics in Cologne, He reports:

> The following year, we began to construct a new secret rocket research center—it was secret from the rest of the world and also secret from the German Army. The Army had a monopoly on rocket work and the Air Force could not get official permission to conduct any research.[65]

By late 1938 he had begun a wind-tunnel test series to determine and refine the flight characteristics of his "Silver Bird" spaceplane design. A year later the outbreak of war threatened the entire project, because the Luftwaffe was determined to focus its development efforts on projects that would have immediate military applications. As Sänger's co-worker, who later became his wife, put it, the " 'Rocket Spaceplane' had to have a new covering if it wished to survive and had to serve a rocket-bomber project." And so the spacecraft was "camoflaged as 'rocket bomber' during the storm of the Second World War."[66]

Sänger wrote a new proposal, presenting his Silver Bird in military dress, but it was rejected in early 1942. He was able to get it published as a top-secret document for dissemination within the closed circle of German aviation in 1944. This monograph described the Silver Bird as an *antipodal bomber,* a rocket aircraft capable of flying halfway around the Earth at high speed, dropping bombs on any target along the way. The device might be feasible today and bears some superficial resemblance to the Space Shuttle currently under development in the United States, but the long rocket aircraft development programs of the past 30 years suggest that the idea of the Silver Bird was premature in the early 1940s. Certainly, Germany came nowhere near developing such a weapon, and it could hardly have been effective. It required a huge launch ramp to boost it above the speed of sound. Whatever difficulties there might have been in building this 3-kilometer catapult, a strong enemy could easily have disrupted its operations with conventional air attacks. The limited payload of explosives deliverable to the target would hardly have been worth the effort, although Sänger went to great lengths to try to give the opposite, more optimistic impression.[67]

As late as 1958 Sänger asserted his antipodal bomber had great advantages over other weapons, even over the ICBM.[68] After 1945 the Silver Bird could be imagined as a bearer of atomic bombs. Basically it is not a bomber, however, but a spacecraft. In his German career

Sänger attempted, like von Braun, to exploit the military and obtain support for his space project through misrepresenting it as a weapon.

CASE 9

PURPOSE:	Holding together the core of the Peenemünde team through the German defeat to continue work on spaceflight
SPACEMAN:	Von Braun
PATRON:	United States
OPPONENT:	Japan (briefly), U.S.S.R.
PATRON'S INFERIORITY:	General desire not to be again unprepared for war; U.S.A. not at the time actually inferior and therefore giving only very weak support

On January 27, 1945, after two weeks of rapid advance, the Russian Army was within 150 kilometers of Peenemünde, and von Braun realized that the German Space Program was at an end.[69] He called together his closest associates to discuss the situation. Dornberger had long since left for other duty, and von Braun had personally achieved an almost feudal command over his people. His authority was in great measure charismatic and was strengthened rather than weakened by the decline of the Third Reich. He put a question to his colleagues: Should they remain at Peenemünde to be killed or captured by the Russians, or should they travel south to be captured by the Americans? Von Braun preferred the latter course, and the others agreed. The cunning they had learned in 13 years of exploiting the German Army and the Nazis came to their aid. Among the various contradictory orders received over the previous weeks were some directing them to move to safer parts of Germany and others directing them to hold fast and die as common soldiers. The Peenemünde men decided to follow the orders to move and discarded all others.

Their planned route would take them diagonally across much of the country, and they would have to convince many petty Nazi chieftans along the way to let them pass. They invented a fictitious *Vorhaben zur Besonderen Verwendung* (Project for Special Disposition) and constructed official-looking and impressive red-and-white signs bearing this legend to be carried by all their vehicles. The Peenemünde exodus rivaled that of Moses and his people and might be compared to the migration of a

Teutonic tribe, following their chief in penetrating the West. Five thousand employees, many with families, great quantities of documents and other paraphernalia were transported by ship, train, truck, and car to Bleicherode in the Harz Mountains. Here, at the Mittelwerk factory, many of the operational V-2's had been manufactured. The area was near the so-called National Redoubt, where Germany might make its last stand, and also in the path of the advancing American armies.[70]

As it happened, SS General Kammler, an old enemy of von Braun and his team, was in command of that sector. Kammler had been in charge of concentration camps at one time, and thought it wise to seize the rocket scientists and use them to bargain for good treatment from the Allies. He held von Braun and 500 of his followers in an abandoned army camp, behind barbed wire and under guard. By an intricate series of subterfuges, the team was able to escape from this trap and avoid various hazards set for them by other SS officers.[71]

Dieter Huzel, one of von Braun's lieutenants, successfully carried out his leader's orders to hide several tons of valuable documents, including complete plans for all of Peenemünde's projects. He buried them in a mine shaft, which was then sealed by explosions. The location was kept secret and was to be used in striking a bargain with the Americans.[72]

When the time was right, Wernher sent his English-speaking brother, Magnus, to make contact with the Americans. Charles L. Stewart, the American Army intelligence agent who interrogated the chief rocket scientists, reported: "They had selected the Americans, as they were favorably disposed to this country generally and also because this country was the one most able to provide the resources required for interplanetary travel."[73]

Another member of the Peenemünde team explained the choice of the rocket scientists' new nationality more colorfully: "We despised the French, we were mortally afraid of the Soviets, we did not believe the British could afford us, so that left the Americans."[74] Von Braun himself has said: "My country has lost two world wars. This time I want to stand on the side of the victors."[75]

The Americans and British had fairly aggressive programs under way to appropriate German technical talent, weapons information, and scientific equipment, even acting in violation of agreements with the Russians and each other to accomplish the scientific looting of the country. But their enthusiasm waned. For a time, American officers expected the war against the Japanese might drag on for some time, perhaps long enough to justify enlisting some German aid against them. After the Japanese surrendered, interest dropped, only to be revived as tensions with the Russians increased.[76]

Von Braun quickly wrote out for his American captors a *Survey of Development of Liquid Rockets in Germany and their Future Prospects,* which amounted to an advertisement and prospectus for employment in the United States. He not only described the actual rocket vehicles developed at Peenemünde and their future military applications, but discussed the construction of artificial earth satellites and asserted, "when the art of rockets is developed further, it will be possible to go to other planets, first of all the moon." He predicted:

> ...a well planned development of the art of rockets will have revolutionary consequences in the scientific and military spheres, as in that of civilization generally, much in the same way as the development of aviation has brought revolutionary changes in the last 50 years.[77]

As Klee and Merk have said of von Braun's team: "They were the only ones in the world who understood how to build large rockets."[78] If any nation wanted to build such devices, it would have been well advised to collect the German talent in 1945 and 1946 when it was essentially free. But was any government interested? British interest evaporated immediately. Russian interest did not really develop for two years. American interest was weak. Nonetheless, in *Operation Paperclip,* designed to transport German technical men and machinery to the United States, 118 men of the original Peenemünde team were brought over as a unit, followed by stragglers like Krafft Ehricke and Dornberger. In a valuable study of *Paperclip,* Clarence Lasby reports that 642 specialists in several fields were ultimately appropriated by the United States.[79] Their overall contribution to American science and technology has been great. Lasby reports that by 1960, 126 were listed in *American Men of Science.* A sample of 165 of these men who responded to a questionnaire sent by Lasby had published 30 books, 1260 articles, 1315 unclassified patents, and had filed 734 patent applications.[80]

A hundred captured V-2s arrived in New Orleans by the end of June 1945.[81] Von Braun himself was in Boston by September 19, 1945 and within a few days was near Washington looking over 14 tons of Peenemünde documents.[82] When Dornberger was rescued from a

British prison in 1946 and offered a job by the Americans he joked: "I'm warning you, gentlemen! I have cost Germany billions of marks. You have to reckon with the possibility that you will wind up giving me billions of dollars!"[83]

Until 1950 von Braun's team worked at a variety of modest projects in the United States. They catalogued and interpreted the captured documents. They fired captured V-2s on military and scientific test flights. They undertook various modest design and development projects, and some of them took jobs in industry. Amazingly, von Braun was able to keep together most of the core of his team, transport them to the United States, and get into a position to use the Americans just as he had earlier used his own people.

CASE 10

PURPOSE:	Further development of Peenemünde designs
SPACEMAN:	Helmut Gröttrup
PATRON:	Stalin and the Soviet Union
OPPONENT:	The United States
PATRON'S INFERIORITY:	Russia encicled by American bases and outclassed by American Air Force which has atomic weapons; Russia seen by the world as technologically inferior

In 1945 the Russians seized not only Peenemünde but also the Mittelwerk (Nordhausen) factory that had been constructed under a mountain to produce V-1s and V-2s. Unfortunately for them, both places had been fairly well cleaned out to the benefit of the Americans.[84] Most distressingly, they had failed to capture any of the most important rocket scientists and engineers. To remedy this situation, they immediately attempted to recruit men from the Western Sectors who were being held by capitalist armies under conditions of some privation. As Lasby reports: "the Soviets were not averse to using guile and trickery. A favorite artifice was to lure the dependents of scientists into the Eastern zone in order to hold them as hostages until the scientists appeared".[85]

In time, the Russians were able to recruit enough men to resume V-2 production at the Nordhausen factory, under the direction of Helmut Gröttrup, a junior associate of von Braun's. He assembled an organization to rival the original Peenemünde team but never equal to it in quality.[86] Gröttrup worked under an agreement that assured the Germans they would not be taken away from their homeland, while von Braun's group was forced to migrate to the United States. In violation of this contract, on October 22, 1946 the East-zone Germans were awakened without warning early in the morning and shipped to Russia. By the end of 1946, according to one report, 92 special trains with 6000 German scientists and technicians and 20,000 members of their families passed through Brest-Litovsk into the Soviet Union.[87]

Gröttrup was kept in charge of the main group of experts and installed at Gorodomlya, "a wildish little island" not far from Moscow.[88] According to one of the Germans, Dr. Peter Lertes, "the Russians' aim in the first five years was simply to pump us dry of everything we knew in our field of science."[89]

Their first job was to improve the performance of the V-2, and the first successful firing of a Russian V-2 over a range of nearly 200 miles was made in Kazakhstan on October 30, 1947. The Russians were not satisfied with the work of Gröttrup's team, however, and proceeded to develop their own improved V-2 in a parallel project.[90]

As time passed, the Germans were given work of less and less importance. By the end of 1950 Gröttrup had outlived his usefulness to the rocket projects, and he was demoted to lesser work. At half his previous salary, he designed surface-to-air missiles, did minor ballistics research, then was given the supremely unimportant task of developing propellor-driven iceboats and ultimately computers for lens grinding.[91]

Gröttrup's wife has described her husband as *besessen* (possessed), and he agreed this was a correct judgment.[92] Although we have ample evidence that von Braun was dedicated directly to the achievement of spaceflight from the beginning of his teens, the evidence is not so clear about Gröttrup. He may have been simply attempting to find a splendid and rewarding career for himself and had no dedication to space. Of course the two explanations for his behavior are not necessarily contradictory, and one piece of suggestive evidence exists in support of the claim that he was a dedicated member of the core of the Spaceflight Movement. On March 15, 1944 Gröttrup had been arrested by the Gestapo, along with von Braun and their associate Klaus Riedel whose enthusiasm for spaceflight is commemorated by a Riedel Crater on the Moon.[93] Dornberger was summoned to the office of Field Marshal Keitel, who said:

> Do you know that your "closest colleagues" have stated in company at Zinnowitz that it had never been their intention to make a weapon of war out of the rocket? That they had worked, under pressure from yourself, at the whole business of development only in order to obtain money for their experiments and the confirmation of their theories? That their object all along had been space travel?[94]

Dornberger was able to rescue the trio from imprisonment, and the incident has become legend supporting the high-mindedness of the Peenemünde leaders and tending to absolve them of war guilt. A brief account of the arrest is even written on a display case in the Huntsville, Alabama, space museum, as testimony to the real motives of the spacemen. Actually, the case is not clear, because the incident was one step in Himmler's attempt to take control over Peenemünde and came only after von Braun had refused to cooperate with him.[95] How much Gröttrup was an independent agent or merely a pawn, is also uncertain.

Further, the Soviets do seem to have initiated contact with the German technicians, and we should examine what little is known of Russian intentions and thinking about large rockets in the late 1940s. In April, 1947 the Soviet Government became interested in Eugen Sänger's antipodal rocket bomber designs. Colonel Tokaev, a Russian defector, has provided the only eye-witness account of the events, and although there have been serious doubts about its reliability, we use it here, with caution.[96]

For two years Russian military engineers had been scouring East Germany for useful technical developments, and three of them—Miklashevsky, Moisheev, and Tokaev himself—had independently discovered copies of Sänger's basic monograph.[97] Miklashevsky "raised an immense hullabaloo with the authorities concerning it" and arranged to have the material translated into Russian. Reports by Moisheev and Tokaev were more sober, praising Sänger's work and decrying the backward state of Russian rocketry and aviation, but falling short of proposing that a project be initiated to build a Sänger plane.

The commander of the technical mission in Germany, Lieutenant General Kutsevalov, a man of little training, was enthusiastic and proposed to the Kremlin a crash program to produce the rocket bomber. In a meeting with a group of top Kremlin leaders, including Malenkov and Bulganin, Tokaev presented a restrained report on the feasibility of the design. He urged a general research and development effort, rather than a narrow attempt to build according to Sänger's plans, which he felt were not immediately realizable.[98]

A commission was appointed to investigate the matter. There was evidence that the Western nations might be forging ahead in rocketry. Tokaev told Malenkov and Voznesensky:

> If it be true that the Americans are so greatly concerned with rocket weapons that they have transformed Texas into a vast Peenemünde, as is often said, it is hardly possible that they have overlooked Sänger's plan. They have combed Germany's scientific centres pretty thoroughly. And they have plenty of dollars to spend on pure research work.[99]

Malenkov said he was "not impressed" by the work going on to build Russian V-2s: "...the V-2 is good for 400 kilometers, and no more. And, after all, we have no intention of making war on Poland. Our vital need is for machines which can fly across oceans!"[100]

Sänger's designs seem to have been the only ones then in the hands of the Russian leaders, although the various A-9/A-10 design sketches became known to them about this time.[101]

Almost immediately, Tokaev reports, he was called to an audience with Stalin, attended by top Politbureau members, at which the Sänger project was discussed. Tokaev urged a patient research and development effort. Stalin said:

> Certainly research is necessary. But we still need Sänger planes, and their construction should be our immediate objective...their possession would make it easier for us to talk to the gentleman-shopkeeper, Harry Truman, and keep him pinned down where we want him.[102]

Stalin demanded a vigorous program to develop the rocket plane and said Sänger himself "must be made to volunteer" to help in the effort.[103] The implication was that Sänger should be captured and brought to the U.S.S.R., if he could not be persuaded to come of his own free will. Soviet agents were supposedly sent to Western Europe but were unable to locate him, and this plan was shortly dropped.[104]

Under Sergei Korolyov and other former members of the Russian spaceflight groups, native Russian ICBMs were developed and applied to satellite launch tasks. But the advocacy and technical work of Germans such as Gröttrup and Sänger contributed greatly to the initial stage of the Russian program. The Peenemünde successes were a necessary prelude to Russian work and greatly channeled Russian thinking.

CASE 11

PURPOSE:	Development of a liquid-fuel rocket one step beyond the V-2 capable of launching a minimum satellite: Redstone/Jupiter-C
SPACEMAN:	Von Braun
PATRON:	U.S. Army
OPPONENT:	Communist armies
PATRON'S INFERIORITY:	General competition with socialist nations; interest directly stimulated by Korean War and the perception that mass Communist armies must be countered by advanced technology

American interest in the German rocket scientists was weak until stimulated by "increasing Soviet intransigence" in early 1946.[105] In the light of rocket development since then, the American appropriation of this German talent was an excellent idea. According to one estimate, the total cash value of the information, materials, and expertise brought over from Peenemünde and Nordhausen was $750,000,000.[106] Of course it was only worth a cent because it did in fact become the basis for rocket programs carried successfully through to completion. That the long-range rocket would ever be a valuable weapon delivery system was not at all obvious in the late 1940s. Operation Paperclip brought to the United States not only a valuable technical resource but also a *mental set* defining the ideal rocket designs and a collection of values in which spaceflight was seen as the prime use of such vehicles, and military purposes as secondary. Although there had been some work on liquid-fuel rockets in this country, all military rockets were then solid-fueled. The V-2 and the team that built it established the more spaceworthy liquid-fuel engine as the standard.

According to a perhaps apocryphal story still going the rounds in spaceflight circles, postwar American rocket designers (and their Soviet counterparts) slavishly copied the V-2 in every detail. One day, the story goes, von Braun was shown one of the American designs which featured squared-off tail fins like those of the V-2. Why, the German asked, do you make the tail fins like this? The American replied that it must be the proper shape for tail fins, because the V-2 had four of them like that. The German laughed and replied that the V-2's fins had to be squared off close in to the body of the missile so it could fit through German railway tunnels.

In fact, American rocket engineers departed immediately from the specifics of V-2 designs, but the German work was extremely influential. Writing about the early development of the American Atlas ICBM, which is still used for some space launches, although it has long since been discarded for military purposes, John L. Chapman said:

> The German V-2 was the real spur. After the war, the Army Air Forces shipped three V-2 engines from Germany to North American Aviation for study. One thing led to another, and within a year North American was in the rocket-propulsion business....[107]

In 1949, pursuing a modest rocket development program, the U.S. Army decided to move many of the members of von Braun's team to the Redstone Arsenal in Huntsville, Alabama, where many of them still live. One of the projects under way was Hermes, actually a cluster of design ideas including a further development of Germany's *Wasserfall* antiaircraft rocket.

The invasion of South Korea on July 25, 1950 stimulated an expanded effort, and in July the Army Chief of Ordnance initiated the Redstone rocket project.[108]

> In mid-1950, the Army turned to the development of long-range rockets. Ordnance called upon the Peenemünde emigres, most of whom by this time had become naturalized American citizens, to make a feasibility study of a mobile, 500-mile rocket, which troops could launch in the field. The mission reflected developments in atomic-bomb technology, which suggested the practicality of mating a nuclear warhead to a rocket. The von Braun team was back in the war-rocket business.[109]

Work on the Redstone allowed von Braun's team to develop further their technological ideas, but the size of the Redstone gave it a special advantage over the V-2. Unlike the V-2, the Redstone was just powerful enough to serve as the first stage of a minimum satellite launcher.[110]

By the early 1950s, von Braun and his associates were disappointed by the relatively cool reception given their spaceflight dreams by the American military. Although satellite projects had been considered by the government immediately after the war, no progress had been made since then.

Von Braun took his case directly to the American people. In March of 1952 the first of a series of popular articles outlining the rocket enthusiasts' projected conquest of space appeared in *Collier's* magazine, and "had great influence in creating space mindedness throughout the United States."[111] Although the articles were based on the most detailed theoretical and engineering studies, they were presented in a brilliantly popularized style, decorated and made more real by stunning color paintings of exactly how the vehicles would look in flight, of how the conquered moon would appear, and of enticing Martian vistas. These articles were reprinted in expanded form in a series of popular books, and several other authors brought out their own versions.[112]

Not only did the *Collier's* articles stimulate public enthusiasm and such commercial responses as a Walt Disney movie, but they served to establish a network of communications between several groups of American space travel intellectuals. Because of the articles, von Braun first came into contact with S. Fred Singer, who with James A. van Allen and others was most influential in creating the International Geophysical Year satellite project.[113] However, this burst of popular interest did not result in any public movement to secure funds for an independent space program. The IGY satellite was given low priority.[114]

When the time came for the "Stewart Committee" to decide between rival earth-satellite proposals for the IGY, there were two competitors to the German team: the Air Force and the Navy. Each of the three services wanted to be responsible for the World's first artificial satellite; the Navy was chosen. The Navy's Vanguard, to be derived from the successful Viking, an American research rocket inspired by the V-2, seemed more feasible than the Army designs, not only in its expected ability to loft payloads into orbit but also in potential for growth. To accept the erroneous public opinion that Vanguard was a failure would be a great mistake. It achieved its original objectives despite many problems and intermittent financial support.[115] The very successful Agena satellite stage is a descendant of Vanguard's second stage. The Air Force proposal was rejected because the Atlas ICBM project might be delayed if Atlas facilities and boosters were taken over by the satellite mission. The Air Force Discoverer Satellites, prototype spy satellites, would undoubtedly have been orbited around 1960 even if no other satelites had been launched by the United States or U.S.S.R.

> The chairman of the Committee admitted privately that some of the ad hoc Group disliked the idea of using a booster that was a modification of a Nazi Vengeance missile developed by German engineers; an American IGY satellite launcher should be an American product. But, Stewart added, the line of reasoning had had little bearing on the majority's decision.[116]

Even though their proposal had been turned down, the von Braun team continued to work on their satellite, Project Orbiter. The Jupiter-C booster that was the basis of their design was not a true Jupiter but an updated version of the Redstone. It bore the name of its offspring for the cunning reason that Jupiter missiles had a high launch priority at Cape Canaveral, while Redstone, which already had been tested thoroughly, had a low priority in a time when launch facilities were extremely limited. Supposedly, the Jupiter-C was built to test nose cones and reentry procedures. Was it pure accident, however, that the vehicle was capable of putting its fourth stage into orbit?

Some military leaders, such as General Gavin, actually suspected von Braun would try to orbit a surprise satellite. On September 21, 1956 von Braun's team launched a Jupiter-C which delivered its final stage a distance of 5500 kilometers. Von Braun complained that he could have launched a satellite that day, or soon after. General Medaris had been ordered to make sure the fourth stage of von Braun's satellite launcher would not achieve orbit. Instead of carrying a live solid-fuel motor, it was therefore loaded with sand.[117]

General Medaris, the American general who had replaced Dornberger as von Braun's patron, reminded his superiors that a second Jupiter-C in satellite configuration was ready and able to be launched on short notice. He testified:

> At various times during this period we suggested informally and verbally that if they really wanted a satellite we could use the backup missile. In various languages, our fingers were slapped and we were told to mind our own business, that Vanguard was going to take care of the satellite program.[118]

There was widespread conviction on the part of the Vanguard leaders

> ...that the Army missile team had jumped the gun by preparing for a satellite shot years before getting authorization to do so. Apprised of this frequently advanced charge years later, former President Eisenhower expressed surprise, saying, "but that would have been a court martial offense!"[119]

After Sputniks I and II and the first Vanguard failures, von Braun was finally given permission to launch a satellite using his Redstone/Jupiter-C. Explorer-I achieved orbit January 31, 1958, the first American satellite, boosted by a missile that was in effect an advanced V-2 designed by the German team. Russian successes in space, and the apparent incapacity of our own rockets, gave a special boost to the fortunes of von Braun and his team. For the 10 years after Sputnik I, a "space race" was in heated progress between the U.S.S.R. and the United States, and the rivalries between the two nations were effectively exploited by the Spaceflight Movement. As NASA administrator T. Keith Glennan had said: "When, as in the space program, the United States is made to look second-best, there is an outpouring of public opinion in this country demanding that we regain first place."[120]

CASE 12

PURPOSE:	Continued development of small space boosters
SPACEMAN:	Von Braun and his team
PATRON:	U.S. Army
OPPONENT:	U.S. Air Force and Soviet Union
PATRON'S INFERIORITY:	Air Force Thor and Atlas projects under development; fear of Russian Army rocket forces

America's Intermediate Range Ballistic Missiles (IRBMs) are often falsely imagined to have been the natural transition stage from the modest Redstone to the full ICBMs, such as Atlas, Titan, and Minuteman. Actually, the Thor IRBM was developed only after the Atlas was well under way, using some parts developed for Atlas, as a means of quickly producing a stop-gap missile to hold the line until Atlas itself was ready.[121] Stationed at British bases, Thor could parry Russian

ICBMs effectively until a credible American ICBM force had been developed:

> In the 1954-55 period...the top levels of the United States government, such as the National Security Council, realized there would be a gap some five years ahead. Intelligence reports indicated Russia would have a dangerous number of nuclear-warhead ballistic missiles operational by 1960. Engineering reports said we could not possibly have a comparable number of InterContinental Ballistic Missiles (ICBMs) to cancel out that threat.[122]

Von Braun's Jupiter missile project was begun apparently slightly later than the Thor, as IRBM-2, a back-up missile, just as the Titan ICBM was to be the back-up for Atlas.[123] After completion, Jupiter was used for a few modest deep-space and satellite missions, in a configuration known as Juno.

The projected missile gap never actually occurred, or if one can discern a "gap" in the late 1950s, it simply did not matter.[124] Throughout this period the superiority of American strategic bomber forces remained the significant atomic fact. Even as late as the mid-1970s, American deterrent strategy, the so-called "Triad" defense, was based on the assumption that the Soviet Union could be effectively destroyed by American bomber forces alone, even if not a single missile was fired.[125] The "missile gap" existed only as a psychological fact in the minds of military planners, congressmen, and the public.

General Medaris, von Braun's U.S. Army patron, was intensely involved in competition with the Air Force and has argued that the myth of a "bomber gap" in the 1950s was a false belief fostered by Air Force leaders to their advantage, but the disadvantage of the nation and of the Army.[126] He believed the competition between the Army's Jupiter and the Air Force's Thor was of great importance as well: "It had become a very competitive struggle between the Air Force and the Army....whoever was assigned this particular weapon [IRBM] would automatically take over the intermediate battle area."[127]

On November 6, 1956 Secretary of Defense Wilson gave the Air Force the prime strategic missile role and limited the range of new Army rockets to 200 miles.[128] By July, 1960 von Braun's team had been transferred from the then stymied Army rocket program to the new and growing National Aeronautics and Space Administration.[129]

CASE 13

PURPOSE: Development of an efficient launch vehicle capable of orbiting large payloads for effective exploitation of near-earth space

SPACEMAN: Von Braun

PATRON: U.S. Army, then NASA

OPPONENT: U.S.S.R.

PATRON'S INFERIORITY: *Sputnik* gives U.S.S.R. prestige advantage; false fear that the Russians might develop superlarge military rockets and space stations

In 1957 von Braun's Huntsville team independently submitted a document to the Department of Defense titled: "Proposal for a National Integrated Missile and Space Vehicle Development Program."[130] Included were first specifications for the Saturn I launch vehicle. This rocket is a direct descendant of the Redstone and Jupiter; in fact, the first stage of Saturn I can be conceptualized as a Jupiter surrounded by a cluster of Redstones.

In truth, no military mission for Saturn I has developed. Air Force Titan vehicles have proved sufficient for even the largest spy satellites. But in the late 1950s gung-ho military men could be deluded about the natural course of military space exploitation.[131] Military bases on the moon and even transport of troops by rocket seemed plausible.[132] Medaris mistakenly claimed:

> It was perfectly clear to me—and had been clear to von Braun for years—that the first nation to establish a permanent, manned space station would have taken a giant step toward domination of the whole planet.[133]

Von Braun wanted such a space station as a major part of his overall plan for interplanetary travel. Space vehicles could be refueled at such a station, and deep-space craft could be assembled from a space station in orbit. Saturn I could transport the parts, if somewhat inefficiently, and has in fact proven its worth as a crew transport system by taking three groups of astronauts to the Skylab station.[134] However, the kind of station envisioned by von Braun could have only the most limited military value; modern spy satellite systems use a number of smaller satellites in various orbits, thereby covering larger land areas than could be managed by a single station.[135] Furthermore, a large manned station would be a "sitting duck" for Russian antisatellite missiles. Von Braun wanted to sell spaceflight and throughout the 1950s oversold its military value.

As usual, the Peenemünde-Huntsville team resorted to every possible subterfuge. They worked on a joint project with RCA to develop a camera for spy satellites, although the Air Force had been given this role. "We had no mission to put up satellites....The RCA project was therefore justified under the heading of a damage assessment device....[136]

By the time Saturn I actually flew, it was under the auspices of NASA. The last flight of a Saturn I was made July 1975, when it launched the *Apollo* vehicle participating in the joint Russian–American *Apollo Soyuz* Test Project.

CASE 14:

PURPOSE: Development of superbooster suitable for lunar and space station missions

SPACEMAN: Von Braun and many others in NASA

PATRON: The Kennedy-Johnson Administration

OPPONENT: The Soviet Union, the Republican Party

PATRON'S INFERIORITY: American spacefaring inferiority as evidenced by Russian *Sputniks* and Gagarin's orbital flight; Kennedy's "New Frontier" in trouble (partly because of the Bay of Pigs fiasco) and needs new visionary boost to regain credibility

Misled by its own romanticism and by the partisan advice of "the best and the brightest,"[137] the Kennedy Administration started the United States on two expensive journeys. A course was set for the Moon, and American power moved blindly into Vietnam. In both cases real experts were ignored—"old China hands" who really understood Southeast Asia and astronomers who knew the Moon. Instead, the military and the spacemen were able to sell false solutions and encourage unreasonable optimism. In both cases there was a failure to bring new understanding to new situations and a tendency to misperceive them as identical to earlier situations. Vietnam was falsely perceived as a new Korea. The Moon was mistaken for a piece of strategic real estate, for the New Frontier (or the Old Frontier), and for a badge of status to be won either by the United States or the Soviet Union. If the Russians had reached the Moon first (they may have come near to flying around it—we do not know for sure—but had none of the hardware necessary for a landing), perhaps the Red Star would have risen higher in the nebulous heavens of international prestige. America does not seem to have profited much as the victor.

Both the Moon and Vietnam were wars, or the hottest fragments of the same Cold War. Both were won, and both in the end were lost. Americans went to both places, made news, spent billions of dollars, then came home. At least no natives were killed on the Moon. In the naive terms of the Cold War, both outposts have now been abandoned.

John Kennedy took office after a very close election and with the strong feeling that he had to prove himself to the American people and to the Russian leadership. The second youngest U.S. Chief Executive in history, he had campaigned on a platform of almost adolescent idealism and had to demonstrate his maturity, his toughness, and his decisiveness. Before the election he had followed the lead of his running mate, Lyndon Johnson, and accused the Republicans of permitting the Russians to forge way ahead in missile development. When he took office he discovered this was not true. The realities of government quickly stalled his program to remake the nation.

Kennedy was inaugurated January 20, 1961. On April 12 Yuri Gagarin, the first Russian cosmonaut, orbited Earth. Only a week later Kennedy had to admit the total failure of the invasion of Cuba at the Bay of Pigs. Vice President Johnson had been given the job of overseeing the space program. Because vice presidents are usually underworked and the space program was not yet under some other political administrator, Johnson was able to make space his private project and a source of at least a little personal prestige. Vice presidents since then have also boosted space as a way of building themselves up; Spiro Agnew was the most vocal advocate of a project to land men on Mars, and Hubert Humphrey has been honored by the naming of a bridge at Cape Canaveral after him. Over the two weeks following the Bay of Pigs, Kennedy and Johnson conferred with their space advisers, including von Braun, about the possibility of achieving some triumph in space ahead of the Russians.

From the early 1950s von Braun had been trying to sell the American government on a moon program. On April 29, 1961, in response to a request for an outline of possible future space goals, he sent Vice President Johnson a memorandum that included an energetic and reasoned argument that the ideal next goal was a manned circumnavigation of the moon and, perhaps, a landing. Among the criteria supporting the political value of such a mission was the assessment that these were the nearest goals in space that the United States could hope to achieve before the Russians.[138] On May 5, the suborbital flight of Alan Shepard in a Mercury capsule showed that the United States could develop a spacefaring capability.

Three weeks later Kennedy made the public announcement that took Armstrong to the Moon eight years later. At the Kennedy Spaceport visitor center in Florida there is a large bronze plaque bearing the President's head in profile and the following legend.

> *I believe this nation should commit itself to achieving the goal before this decade is out of landing a man on the Moon and returning him safely to Earth*
>
> *John F. Kennedy*
> *MAY 25, 1961*

The Peenemünde team was by no means the only source of support for the program that came to be known as Apollo, and the bulk of the engineering work was in fact done by native Americans. However, the essentials of the Saturn V launch rocket were designed precisely by von Braun's team; it could not have been built without the long experience with large, liquid-fuel rockets beginning with the V-2. The political and historical processes that led to the landing of *Apollo 11* have been analyzed at length by a number of competent authors most notably Vernon van Dyke[139] and John M. Logsdon.[140] Here, I have only been able to describe the major events in outline. Scholarship on the period beginning just before *Sputnik I* and ending with *Apollo* is reasonably complete, and the major events clearly fit our model.

Progressively, the competition for supremacy in space became autonomous, and the elite of the Movement no longer always had to initiate discussions with government officials over future projects. Rather, their role has become that of helmsmen, whenever possible steering the line of rocket development in the directions most suitable for spaceflight.

The Skylab project marks a transition point. Although it was a fairly expensive undertaking, it used one Saturn V and three Saturn I launch vehicles left over from the Apollo moon program. Competition with the Russians remained an important motivation, but there was considerable propaganda issued by NASA selling Skylab as a scientific venture promising economic returns. There was a growing sense that the United States had trounced the Soviet Union in the space race and that further exertions would have to serve more pedestrian motives. The Russians themselves had developed a large booster rocket even more powerful than the Saturn V, but all three launch attempts were disastrous failures.[14] The Russian competition to Skylab, Salyut, is a more modest space laboratory, and the Salyut program has been marked by several failures, including the *Soyuz 11* accident in which three cosmonauts returning from a Salyut were killed.[142] This situation might stimulate the Russians to further space investments, but they may feel the necessary expenditures would be so huge as to be foolhardy. There is some evidence suggesting that they have not given up on plans to develop a large manned space station, but the future course of the Soviet space program is unclear. In any case, from Skylab on, the model we developed here, which well explained mobilization of resources for the earlier stages in space development, probably no longer applies.

CASE 15

PURPOSE:	Development of a Space Shuttle orbital vehicle for the consistent exploitation of near-earth space
SPACEMEN:	Dornberger, Ehricke, NASA, American aerospace corporations
PATRON:	United States Government
OPPONENT:	None
REASON FOR INVESTMENT:	No inferiority vis-à-vis another power; rational calculation that launch costs would be reduced by more efficient booster systems; NASA has become an entrenched governmental institution, and is able to command fairly constant appropriations, so long as plausible projects can be sold to Congress.

With the Space Shuttle, the Spaceflight Movement in America has ceased to be a movement and has become institutionalized as a part of the standard government-industrial structure. The Spaceflight Movement has matured, and in succeeding, has lost its power for revolutionary growth. At the present time there are neither great technical opportunities nor intense social pressures that can produce a new wave of the Movement leading, for example, to planetary colonization or some other great step. A second wave of the Movement is of course possible, but the conditions necessary for it do not seem to exist now.

Spaceplanes capable of flying to and from orbit without having to discard expensive booster stages are not a new idea. Valier and Sänger were both active proponents of this approach, but made their sketches rather prematurely, before general rocket technology had reached the point where workable vehicles could be built. In the early 1950s Peenemünde men Krafft Ehricke and Walter Dornberger designed a two-stage rocket aircraft in which both stages would be reusable.[143] For a time, it looked as if the Air Force would invest in a modified spaceplane called *Dynasoar,* short for dynamic soaring:

> In 1951 Dornberger had advanced a plan for what later became Dynasoar, the experimental vehicle for manned space flight. Dornberger acknowledged that he had had to make exactly 678 presentations before various committees and independent groups in authority before he could get Dynasoar off the drawing board.[144]

Dynasoar was cancelled, but on January 5, 1972 President Nixon announced:

> I have decided today that the United States should proceed at once with the development of an entirely new type of space transportation system designed to help transform the space frontier of the 1970's into familiar territory, easily accessible for human endeavor in the 1980's and '90's.
>
> This system will centre on a space vehicle that can shuttle repeatedly from Earth to orbit and back. It will revolutionize transportation into near space, by routinizing it. It will take the astronomical costs out of astronautics. In short, it will go a long way toward delivering rich benefits of practical space utilization and the valuable spinoffs from space efforts into the daily lives of Americans and all people.[145]

While the earlier space vehicles carried glorious names like Mercury and Apollo, this new "space transportation system" will use the name

Space Shuttle for its prosaic workhouse capable of making as many as 100 round trips to orbit. Although Ehricke is affiliated with North American Rockwell, the prime contractor for the Shuttle, the ship will be a thoroughly American creation, building on that corporation's experience with the X-15 rocket plane. The basic concepts of the Shuttle derive from work by members of the German branch of the Spaceflight Movement, but the German Space Program is now finished. Wernher von Braun himself has entered partial retirement and lives, appropriately enough, in Germantown, Maryland. Important NASA posts are still held by alumni of Peenemünde, such as Eberhard Rees and Kurt Debus, but the American space program, like the Russian, has been thoroughly nationalized.

Our exploitation model no longer holds, and whether there are any new projects spacemen can sell now to military men is not clear. The solid-fuel ICBM seems effectively perfected. This does not mean, of course, that there will be no spinoffs from military technology into space applications. The prototype attitude control engines for the Space Shuttle, the small rocket engines that adjust the orientation of the craft in space, are being developed from the motors built for the fourth stage of the Minuteman III ICBM.[146]

As NASA historian Eugene Emme says, the development of the Space Shuttle marks the end of the "heroic phase of space flight." The revolution is finished, and an era of normal technological advance in space has begun.[147]

Chapter 4 / TECHNOLOGY ASSESSMENT OF THE PEENEMÜNDE ROCKETS

The social impact of a technological development cannot be understood without a thorough technical analysis of the development itself and alternate developments that could have taken its place. Although this principle seems obvious once it has been stated, it has often been violated by writers in the field. Technical histories tend to stress technical details; sociological studies tend to stress sociological relationships. All too rarely do authors enter that crucial territory that is both technological and sociological. In this chapter we analyze the technological significance of the Peenemünde developments, stressing the actual capabilities of the rockets but without losing sight of their social meaning.

Comparison of Liquid-Fuel and Solid-Fuel Rockets

The V-2 stood 14.3 meters tall, weighed 4000 kilograms empty and 12,800 kilograms fueled. It burned 3800 kilograms of alcohol with 5000 kilograms of oxygen stored in liquid form.[1] When fired without warhead for extreme range it could achieve over 300 kilometers. On December 17, 1946 a captured V-2 fired from White Sands Proving Ground in New Mexico was able to reach an altitude of 187 kilometers.[2]

The major impact of the V-2 was as a demonstration that large liquid-fueled rockets were feasible and might be employed as weapons.

One indication of how important this demonstration was is found in the fact that various informed British Intelligence officers were extremely reluctant to accept the reports they were receiving out of Nazi Germany about long-range rockets and refused to believe that the missiles were liquid fueled until they had specimens of them. Although modest but successful tests of gasoline/liquid oxygen rocket engines had been conducted in Britain and the United States and the *VfR* research had been well publicized, the experts' belief that Peenemünde's rockets used solid fuel persisted until late September 1943.[3]

All three great pioneers, Tsiolkovsky, Goddard and Oberth, realized the superiority of liquid-fuel rockets over solid-fuel rockets for space purposes. This discovery and the work that each did in developing the concept were precisely the accomplishments that made them pioneers and encouraged later decades to consider them great. All three urged the use of liquid oxygen with either liquid hydrogen or more common liquids such as gasoline or alcohol instead of gunpowder or some more advanced development from conventional fireworks.

Certain liquid propellants contained more available energy than solid propellants for the same weight. In 1936 John Shesta of the American Rocket Society compared several propellants in terms of how many BTUs were liberated per pound. The black powder of the typical fireworks rocket gave off only 1000 BTUs; weapons-grade smokeless powder achieved only 1870; methyl alcohol burned with oxygen produced 3030; hydrogen burned with oxygen produced 5760.[4]

Another way of looking at the same differences is in terms of the theoretical exhaust velocity that could be achieved with different propellant systems. The typical modern (1970s) solid fuel such as used in military rockets could in principle achieve 2250 meters per second. The fuel combination Goddard actually used, gasoline and oxygen, could reach 2750 meters per second. Hydrogen burned with oxygen achieves 4200 meters per second.[5]

Today the comparisons are usually made in terms of *specific impulse,* a measure unfortunately confusing to laymen because the unit is called *seconds.* As Peenemünde's Huzel points out, "Obviously, the expression does not denote a time, but rather a magnitude akin to efficiency."[6] Specific impulse can be calculated by dividing the exhaust velocity by the acceleration of gravity at the earth's surface,[7] which in the metric system is very close to 10. Typical modern solid propellants achieve a specific impulse around 230; the best modern solids reach 285 to 290, and there is some hope that future solids might get as high as 310 seconds.[8] This would still be far below the figure for hydrogen and oxygen, which is 420.[9] In simple terms, the best liquid fuels are more powerful than the best solid fuels. Because launching a large payload at spaceflight velocities is difficult, liquid fuels obviously have a great advantage. Because ICBM velocities are considerably lower, the increased efficiency of liquid fuels may not be an important consideration for them.

The earlier pioneers also thought that a higher *mass ratio* could be achieved using liquid-fuel engines. The mass ratio is calculated by dividing the takeoff, fully fueled mass of the vehicle by the final, empty mass. A "high-performance" solid-fuel rocket tested in 1730 by German military officers had a mass ratio of 1.33. More than 200 years later some World War II bombardment rockets had mass ratios no higher.[10] British Intelligence in 1943 was of several minds about the technical aspects of the German long-range rockets they had been hearing about from their agents. Using a quotient, *alpha,* related to mass ratio, they attempted to discover if the V-2 could have a significant range or payload. Alpha was simply the mass of fuel carried by a rocket divided by its total fueled mass—that is, the percentage of the rocket's mass represented by fuel. The best alpha achieved by cordite solid-fuel rockets in Britain was 0.25. That is, the fuel was a quarter of the total weight, and the mass ratio was again close to 1.33.[11] One British guess for the V-2 configuration thought it might be a 10-ton rocket of which only 4 tons would be a solid propellant, and the remaining 6 tons the heavy casing required to confine the fuel's explosive force.[12]

In a classic solid-fuel rocket the entire length of the tube acts as the engine. The fuel must burn where it is, and so the entire structure that contains it must be able to sustain the highest pressure. In a liquid-fuel rocket only the relatively small motor itself, into which the liquids are pumped for burning, must be able to sustain high pressure. The fuel tanks can be lightly built. Even after the British had begun to realize that the V-2 was liquid fueled, they remained conservative in their estimates, in one case assuming the highest conceivable alpha was 0.58, implying a mass ratio of 2.4. The actual alpha of the V-2 was 0.64, giving a mass ratio around 3.[13] The final velocity achievable by a rocket is highly sensitive to mass ratio; so the difference in performance between the V-2, and the hypothetically comparable solid-fuel rockets is extremely great. The use of contemporary casing materials and improved design have significantly bettered the mass ratios attainable with solid fuels.[14] However, liquid-fuel mass ratios have also improved, approaching 10, for example, in the case of the Atlas ICBM.

In all of these theoretical comparisons liquid fuel wins out over solid fuel. Why do almost all current United States military rockets use solid propellants? Let's examine briefly the development of the first solid-fuel IRBM, the Polaris.

In 1955 the U.S. Defense Department decided to have the Army develop an IRBM which would also be available for use by the Navy. The Army called upon the von Braun team which had already produced the Redstone, an offspring of the V-2, and the liquid-fuel Jupiter IRBM program got under way.

The Navy was not pleased. First, they wanted their own missile, not a hand-me-down from the rival Army. But the Navy's chief complaint was that the Jupiter's liquid fuels would be difficult to handle at sea.[15] In the late days of World War II the Germans had experimented with a scheme for towing V-2's in special cannisters behind submarines. Their plan was to bring the missiles close to the American shore and fire them at seaboard cities such as New York, Boston, or even Washington.[16] The U.S. Navy wanted an IRBM that could be fired directly from ships, not awkwardly towed about.

In 1947 the U.S. Navy fired a V-2 from the deck of the aircraft carrier Midway; the rocket exploded at an altitude of 6 miles, giving some pause to officers who considered the possibility of an explosion 6 miles earlier.[17] In May 1950 the Navy had fired one of its Viking research liquid-fuel rockets from the deck of the ship Norton Sound but had only succeeded in doing so after a number of delays.[18] For one thing, the weather had to be extremely good. The sea had to be calm and the ship's deck stable before the rocket could be fueled without prohibitive risk. One could not count on such ideal conditions in wartime!

In considering the hazards of Jupiter, the Navy also remembered the results of Operation Pushover. In this simple experiment a fully fueled V-2 was pushed over while sitting on a simulated ship deck. The resulting explosion was so devastating that it might have destroyed a real Navy ship.[19] For a while the Navy considered a mammoth solid-fuel rocket with the range and payload of the Jupiter. Then anticipating a further reduction in hydrogen bomb weight and accepting a reduction in missile range, the Navy developed the Polaris.

The Jupiter was used for a number of space launches around 1960 in a configuration called Juno and fathered illustrious descendants Saturn I and Saturn V. The Thor and Atlas liquid-fuel missiles are no longer of any military use but are the chief launch vehicles, in developed forms, for the American space program, along with advanced versions of the liquid-fuel Titan which still plays a minor role as a specialty ICBM prepared to deliver very heavy warheads to a few targets. Since the last Saturn IB launch in 1975 as part of the Apollo-Soyuz Test Project, all large United States orbital vehicles are liquid-fuel rockets aided at launch by solid-fuel, strap-on boosters. The use of solid-fuel boosters is dictated by economic and political considerations. In the long run, larger liquid-fuel rockets might be more economical but would require major initial investments. Strap-on boosters allow existing models of liquid-fuel launch vehicle to grow in capability at relatively low cost and without the necessity of a political battle over a large new project that has to be completed in one big chunk. The only all-solid-fuel American orbital launch vehicle is the smallest, the Scout, capable of putting less than 100 kilograms into orbit. Scout is based on early versions of the Polaris.[20]

For space use the liquid-fuel rocket has three primary advantages. The first, higher performance, has already been considered. The second is more precise control. Liquid-fuel engines can be switched off almost instantly simply by stopping the flow of fuel into the engine. Several liquid-fuel engines, such as the old Agena, have the capability of being restarted in space. The descent-stage engine of the Apollo lunar lander was even capable of extremely fine throttling. The LEM needed to be able to hover over the moon's surface while the crew looked for a landing spot. No solid-fuel engine could be built with such fine responsiveness to control. The third advantage is reusability. The Space Shuttle engines will be used on flight after flight, while military missiles are never returned to the original owner for reuse!

Aside from a few leftover Titan II's, the only current American weapons using liquid-fuel rocket propulsion are the Lance tactical missile and the warhead "bus" of the Minutemen III ICBM.

The Lance has taken storable liquid-fuel technology to a remarkable level. The missile can be stored, completely tanked up on nitric acid oxydizer and hydrazine fuel for periods of *up to 10 years*![21] Lance can carry a variety of warheads, including nerve gas.[22] It is designed for very precise delivery of the warhead to a small target. *Aviation Week* reports:

> During initial boost phase, guidance and control electronics restrain missile's path within a theoretical mile-long "gun barrel." When system accelerometer senses that a pre-set velocity has been attained, boost is terminated. Then a variable-thrust sustainer engine maintains zero-g on the accelerometer, allowing for compensations in air density and other mid-course error sources. At a predetermined time, sustainer cuts off and missile flies free into target.[23]

Such complex behavior is beyond the skill of traditional solid-fuel propulsion systems. More recently *Aviation Week* reported that a terminally guided submissile, TGSM, was under development for use in the Lance warhead. Lance would carry six to nine of these miniature missiles which would be ejected near the target and steer themselves into it. The LTV corporation is working on a new missile prototype incorporating TSGMs:

> In effect, the new missile will be a prototype of a low-cost, solid propelled, fluidic-guided, less accurate version of Lance, which uses electronic guidance and a prepackaged liquid bipropellant rocket motor. In the case of the propulsion system alone, an LTV engineer expects the solid motor will cost only 1/3 that of the liquid motor now in Lance.[24]

The solid-fuel successor to Lance would not, in itself, be so accurate, but this sloppiness would be compensated for by the homing TSGMs.

The Minuteman III ICBM stands less than 20 meters tall, weighs only 35,000 kilograms at launch, yet can hurl its three fusion warheads 11,000 kilometers.[25] The first three stages of the ICBM, the actual boost stages, are solid-fueled, but the highly complex fourth stage "bus" carries three independently targeted reentry vehicles. Using a liquid-fuel engine, it performs intricate and precise maneuvers to place each reentry vehicle in the proper trajectory before releasing it. The relatively crude solid fuel booster stages throw the bus at the target. The refined liquid engine of the bus corrects for some booster error and makes a series of fine adjustments possible for each hydrogen bomb. In this configuration of the Minuteman III we observe the two types of engine showing off their superior talents.[26]

With perfect hindsight, we can say that military *missiles should always have used solid propulsion,* from the earliest days right through to the

Minuteman. A liquid-fuel excursion was made in the history of rocketry because spaceflight enthusiasts such as von Braun guided the militaries of major nations in the direction needed by the Spaceflight Movement. If not for the success of the V-2 and the propaganda impact of its use in World War II, military missiles would never have used liquid propellants.

The military advantages of solid fuels include cost, storability, reliability, and simplicity. These advantages must be forgone by space vehicles to achieve greater power and control. By its striking superiority over contemporary solid-fuel rockets, the V-2 pointed the direct way toward space but led military technology on a detour.

Cost-Effectiveness Analysis of the V-2

Nazi Munitions Minister Albert Speer called the V-2 rocket program "not only our largest but also our least successful project."[27] From the standpoint of German military needs, the V-2 was not *cost-effective.* Reports vary, but according to one, the total investment in the V-2 and related programs was the equivalent of $2,500,000,000![28] For a similar investment the United States acquired the atomic bomb!

Not only did the V-2 consume great quantities of money and scarce materials, but it also monopolized considerable engineering talent that might better have been set to different tasks. Dr. Steinhoff of Peenemünde estimated that one third of the physical scientists and advanced engineers of Germany worked at one time or another on the V-2 project, although many had no idea that the specific tasks assigned them had anything to do with a war rocket.[29] One knowledgeable historian asserts that between 1943 and 1945 a total of 200,000 Germans and captive workers invested their time in the V-2.[30] Speer complains that as late as the first of January 1945, when development of the missile was supposedly complete, fully 2210 Peenemünde engineers and scientists were still employed by the V-2 program, while only 220 were working on the Wasserfall and 135 on the Taifun antiaircraft rockets, which he thought might be more useful.[31]

Speer illustrated his contention that the V-2 was a costly mistake by comparing it to the American B-17 bomber. The aircraft cost six times as much as the rocket but carried nearly three times as much explosive. The most important difference was that the plane could be used many times, while the rocket was always destroyed by a single use. The action radius of the B-17 was 1600 to 3200 kilometers, while the maximum range of the V-2 was 250 kilometers.[32] He neglected to mention that when a B-17 was shot down, the lives of crew members were spent, but then so were the lives of fighter pilots shot down by the B-17. Neither did Speer explain that extremely accurate bombing could be conducted against specific targets by the bomber, while the rocket could not be aimed more accurately than within a radius of a few miles and therefore was suitable for use only as a strategic weapon directed against large population centers and not as a tactical weapon. Speer describes the military context of his dislike for the V-2:

> 49,400 tons of explosive were dropped on Berlin alone, by which 20.9% of the dwellings were seriously damaged or totally destroyed. In order to direct the same quantity against London, we would have had to employ 66,000 great rockets, that is, the entire production of six years.[33]

We can compare the effectiveness of the V-2 against that of its most direct competitor, the V-1 buzz-bomb. This crude, jet-powered winged robot bomb was a joint project of the Luftwaffe; the Argus Motor Company; and the Fieseler Aircraft Works, a maker of light planes. According to Lusar, the V-1 required only 280 man-hours to construct, for a cost of 3500 marks, which he says was one tenth that of the V-2.[34] The warheads of the two V-weapons were almost exactly equal in power, and although the impact of the high-velocity V-2 somewhat augmented the explosive force of the warhead, the rocket tended to bury itself in the ground before going off, thereby containing the blast to some degree.[35]

Willy Ley has calculated that the V-2 launch failure rate was 7.7% for a series of 1027 such vehicles fired between September 1944 and March 1945. Of about 1300 V-2s fired against London, only 518 actually fell to Earth within the London Civil Defense Region.[36] Part of the inaccuracy was caused by the British who labored to give the Germans a false impression of where the rockets were landing, thereby throwing off the aim. Although the V-2 could not be intercepted, the V-1 could be shot down by either fighter aircraft or antiaircraft guns. Only 32% of the V-1s fired against London reached their target, but this minority still succeeded in destroying 24,491 dwellings and rendered 52,293 others uninhabitable.[37] From the figures reported by Ley, we can construct a comparison table (Table 4.1).[38]

TABLE 4.1 COMPARATIVE EFFECTS OF THE V-1 AND V-2

	V-1	*V-2*
Total missiles fired at England	8,079	1340
Number that reached England	7,488	1115
Number that reached London	2,420	518
Number of people killed	5,864	2714
Number of people seriously injured	17,197	6467

David Irving has surveyed the data on the cost of the V-2 and determined that the mass production cost of each of the 5789 rockets manufactured at the Nordhausen factory was 6320 English pounds, figured in 1945 money. To this he adds £350 for warhead and launch costs and about £5,000 development expenses to achieve a fully amortized, total cost per vehicle of £12,000. He says the Royal Aircraft Establishment felt the V-1 buzz-bomb could be built in British factories for around £115 each and cites an average price paid Volkswagen for actual missiles of about £125 each, a close correspondence.[39] Albert Speer has accepted Irving's figures as exact.[40]

In rough terms, V-1s killed twice as many Britons for half the production cost of the V-2s. The V-1 was the result of a very limited development program. Irving's total cost estimate for the V-2 does *not* appear to make provision for any amortization of Peenemünde or the basic rocket research done there. It represents only the money spent developing the V-2 once it had already been created.

By the end of the V-1 attack on London the English had become very successful in shooting the buzz bombs out of the air. It might be argued that this proves the superiority of rockets like the V-2 over cruise missiles like the V-1, but in fact does nothing of the kind. There were many cheap improvements that would have helped great numbers of V-1s get through British defenses—a slightly more complex guidance system, the ability to fly either higher or lower, more fuel capacity to permit a circuitous course, and greater speed. It was the Spaceflight Movement, not technical fact, that convinced the world the ballistic missile was superior to the winged cruise missile. Even today, the two types of robot bomb compete with each other. Both the United States Navy and Air Force have recently announced they are aggressively developing advanced cruise missiles that are the distant descendants of the buzz bomb.

If we want to prove that the V-2 was a military folly, to show that it was not cost-effective in terms of money is not enough. Sometimes the main limitation on the production of a given type of weapons system is not money but what RAND analysts Charles J. Hitch and Roland N. MacKean call "specific constraints," the supplies of raw materials or basic manufactured goods that set a limit on production quite apart from the availability of labor or other components.[41] For example, in February 1944 Albert Speer surveyed German raw materials reserves and the sources of production. He determined that chromium, an essential metal for use in alloys for the munitions industry, was stockpiled in only a 5.6-month supply. The only significant source of this material that remained to the Third Reich was mines in the Balkans; so Speer concluded: "That meant nothing other than that the war would be at an end about ten months after the loss of the Balkans," regardless of how much iron and fighting blood Germany continued to possess.[42]

Peenemünde's General Dornberger has contended that one great advantage of the V-2 was its fuel. Burning alcohol made from potatoes, the V-2 did not drain gasoline away from the Luftwaffe. Historians have often suggested that Germany should have built more jet fighters rather than V-2s, but Dornberger correctly notes that there was not enough fuel for the planes Germany did build.[43] He does not take up the question of whether anybody wanted to eat the potatoes or if some other use could have been made of the alcohol produced from them.

The V-2 of course used liquid oxygen, alcohol, and a moderate quantity of hydrogen peroxide. Dornberger makes no attempt to test the hypothesis that the total amount of gasoline used at all levels in the process of producing these exotic fuels was significant. The fossil fuel used primarily in the German production of liquid oxygen was probably coal. Synthetic gasoline or other fuels suitable for supplanting gasoline in some applications can be manufactured from coal. Commerical plants doing this based on coal gassification had been operating in Germany from the 1920s.[44] Would the investment in liquid oxygen plants have better been made in synthetic gasoline plants? Peenemünde, which had its own rocket factory and liquid oxygen plant, was fueled by coal.[45] The total fuel consumption of a V-1 buzz-bomb was only 150 gallons of low-grade 80-octane gasoline.[46] The V-2 used great quantities of liquid oxygen and alcohol, as we have seen.

The launch catapult for the V-1 seems to have used more hydrogen peroxide than did the pumps of the V-2.

Dornberger called for the air defense of Germany's hydrogenation plants to protect the supply of rocket fuel; Speer urged the defense of similar plants to secure the production of fuel for armored tanks.[47] If Germany could have done either, it could not have done both.

Although the V-2 used alcohol rather than gasoline, that fuel could have been employed in conventional vehicles. Indeed, according to one report, the carburetors of Peenemünde trucks had been modified so that the vehicles could run on rocket fuel![48] In a more modest account of the same thing, Dr. Steinhoff of Peenemünde had asserted his men could arrange to run their vehicles on a mixture of alcohol and gasoline, if necessary.[49]

The Allied air attack on the oil fields at Ploesti in Rumania, May 28 and 29, 1944, cut German petroleum production in half, according to Speer,[50] but a fuel shortage had been felt as early as late-summer of 1942.[51] Had the Peenemünde team been interested in aiding their country's war effort, they might best have used their knowledge to develop German use of methyl alcohol as a general fuel rather than a rocket propellant.

In a recent article on the use of methanol as a gasoline extender, E. E. Wigg pessimistically concludes that such alcohols will not contribute significantly to solving America's current fuel crisis.[52] His key point, however, is that carburetors of the 1970s use too lean a mixture, while older-model cars could indeed conserve gasoline by running on a mixture of gasoline and alcohol. Of course, by today's standards, the vehicles possessed by Germany in the early 1940s were precisely "older models," and the use of alcohol as a gasoline extender in conventional vehicles could have effectively added perhaps 10 to 15% to Germany's gasoline production.

The V-2 has been called *ein Aluminiumblechfresser,* an aluminum-sheet-eater, and according to Irving's data, consumed 2.5% of Germany's monthly aluminum production in 1944.[53] The V-1 was designed especially to avoid such materials bottlenecks[54] and was constructed from low-grade steel.

By any criteria, the V-2 was not a cost-effective weapon. It could not match the performance of much simpler weapons systems, yet drained money, materials, and talent from its sponsors. Propelled by an advanced and expensive engine, it delivered a conventional and insufficient warhead. Despite the apparent advantages of the modern ICBM, the manned bomber still seems able to compete as a delivery system, and cruise missiles (descendants of the V-1) are still being designed. Its warhead is what makes the ICBM an ultimate weapon.

The Possibility of an Atomic Warhead

If a conventional explosive warhead did not make the V-2 into an effective weapon, what about the much touted German atomic bomb? The fact is, although the Germans began the war with at least 3500 tons of uranium compounds and a good number of the best physicists in the world, at the end they had not succeeded in even sustaining a chain reaction in a nuclear reactor, let alone being in a position to produce a bomb.

The community of German physicists seems to have felt considerable ambivalence about building atom weapons. This is interesting, considering the fact that we have encountered no moral resistance among the rocket enthusiasts against the military application of spaceflight technology! Some German scientists in favor of an atomic bomb project made real, if confused, efforts to find support, but there was no one among them the equal of von Braun in either commanding a team or conning the paymasters. Nor was there a motivated institution or other social phenomenon that could be exploited for developing atomic energy. Speer reported that he discussed atom weapons with Hitler in June 1942, but he wanted to avoid the costly waste of money in such a project, just as he was sometimes opposed to the V-2.[55] Further, while rocketry was pursued primarily by good Christian-German researchers, Hitler was convinced that atomics was a branch of abhorred "Jewish Physics."[56]

By the time any groups in Germany were working toward their first atomic pile, von Braun's team had been in business 10 years and in peacetime had consolidated a position from which it could expand in war. The German atomic physicists were competing inefficiently with each other for money and the scarce materials essential for their research and were primarily concerned with theoretical *scientific* discoveries; the united Peenemünde team was after concrete *technical* accomplishments. David Irving has gone so far as to assert that German physics was then on the decline precisely because it had become too theoretical and had "lost the art of experiment."[57] The German atomic leader, Heisenberg, has countered with the charge that American physicists were not theoretical enough.[58]

In comparison with the American situation and with the atomic warhead missile as a standard, German atomic technology lagged behind rocket technology. However, even if it had not, even if the Germans had developed fission bombs equal to the American ones and had prepared them to be used six months earlier, the V-2 could have played no part as a delivery vehicle. The gross warhead weight of the V-2 was 1000 kilograms, of which only 743 kilograms were explosive.[59] The lighter of the two atom bombs dropped on Japan weighed 4000 kilograms.[60] Clearly, the V-2 could not have carried the early atomic bombs. Even the Luftwaffe would have had trouble delivering one; the bomb load of the standard Junkers 88 bomber was a mere 2500 kilograms.[61]

Although a German IRBM capable of carrying such a massive warhead could have been produced by building on the V-2—for example, clustering six V-2 engines to generate 150,000 kilograms of thrust—the development program would have dragged on for a long time. Not only would such an IRBM have to be bigger but, considering the larger number of components, it would also have had to be constructed to much higher standards of quality. When we consider that Germany had completely run out of time and remembering that the Peenemünde team had been reduced to impotence by Himmler in 1944, we can see that a German IRBM carrying an atomic warhead was quite out of the question.

Although there is no evidence an atomic warhead was ever considered for the V-2, on October 15, 1942 Major Leo Zanssen of Peenemünde contracted for the *Forschungsanstalt der Deutschen Reichspost* to begin "the investigation of the possibility of the exploitation of atomic decay and chain reaction to rocket propulsion."[62] Nothing came of this. Dr. Walter Thiel, leading engine specialist at Peenemünde who had urged the exploration of atomic propulsion, had acted as liaison with atom scientist Werner Heisenberg. Thiel was killed in the British bombing raid on the rocket base. Later, when Zanssen was placed in charge of the military forces at the base, he seems to have lost interest in the project.[63] Zanssen himself was fired from his command on June 1, 1944. SS intrigue reduced Peenemünde from a center of technological revolutionism to the conservative status of a private corporation.[64]

The Question of Poison Gas

Although the Germans possessed no explosive powerful enough to transform the spaceship V-2 into a truly effective weapon, their arsenals were rich with great quantities of poison gas which easily could have replaced the warhead of the missile. According to Cookson and Nottingham, Germany had stockpiled 12,000 tons of Tabun, the first nerve gas, before the end of the war. Production of Sarin, an even more effective nerve gas, had begun.[65]

A standard way to express the strength of various poison gases is through the concentration in air required to kill 50% of the people exposed to it for just one minute. If the concentration of Tabun in the air is 150 milligrams per cubic meter, inhalation for one minute will cause death to half the exposed humans. Of course a number of natural details are ignored or held constant in such an estimate. If the people exert themselves and breathe heavily, they will take in more Tabun and be more liable to die. If they are ill or aged, less gas will be required. The same effect is achieved by a concentration of only 70 milligrams per cubic meter of the more deadly Sarin. In ordinary terms, we are discussing very slight concentrations of what must be described as very deadly substances. The all-around most effective gas of World War I was probably Phosgene, which required a concentration of 3200 milligrams per cubic meter to reach this level of lethality.[66]

Some poison gases, such as chlorine, technically are really gases; others are really a fine spray of liquid droplets. To put Tabun in the warhead of a V-2 requires something more than simply a pressure tank to keep the "gas" in. One would have to use a chemical medium to carry the Tabun, a detonator, perhaps some small plastic "bomblets" to distribute it effectively over an area. In the following calculations, I am assuming that only 50% of each warhead, or 500 kilograms, will actually be the Tabun. I am also assuming that a means could be found to distribute it evenly over the impact area of the missile.

The load of one V-2, 500 kilograms of Tabun, could produce the 150 milligrams per cubic meter concentration in a volume of 3.3×10^6 cubic meters. If we assume that most of the population of London lived and worked within four stories of ground level, it makes sense to ask how much of an area this concentration of Tabun would cover to a depth of 15 meters. The answer is 0.22 square kilometers. The gas-laden V-2 would kill many more people than the conventional V-2 with a ton of explosives in the warhead. It would not even lose all its effectiveness in demolishing buildings, because the fuselage of the rocket would do considerable damage falling at a speed several times that of sound. When V-2s were fired without warheads in tests, they still made a huge dent in the ground.

By the end of January 1945, 560 V-2s had reached England.[67] Our rough estimates suggest that these missiles could have delivered sufficient Tabun to achieve the 150 milligrams per cubic meter concentration in 1850×10^6 cubic meters, or to the depth of 15 meters over an area of 123 square kilometers. The total area of the city of London was about 1800 square kilometers; so they apparently could have delivered only 6.9% of the Tabun required to render London a dead city.[68]

Even this 6.9% would have produced many deaths and great panic, but in fact this figure is probably an underestimate. One hundred fifty milligrams per cubic meter is the concentration required to give the average person a 50% chance of a lethal dose in only one minute of breathing. For purposes of analysis, let us assume the ideal case. Suppose a V-2 perfectly and instantaneously delivers this concentration to an area of 0.22 square kilometers. The Tabun would cover a circle 520 meters in diameter to a depth of 15 meters. On the average, a person in this circle would find himself 190 meters from the center and only 70 meters from the edge. An athlete could run this distance in 10 seconds, and most people could walk it within a minute if there were no obstructions. But London has many obstructions, and the quick breaths taken in running would increase the dose of Tabun ingested. The average person would be indoors and waste time getting out of the building. If, on the average, the impact point were 190 meters away in a crowded city, there might be confusion on where it was and which way to run. Vital time would be lost simply discovering what had happened. Furthermore, sublethal doses of gas can still weaken the organism, endanger the person's general health, and increase the likelihood that a second dose would succeed in killing him.

For concentrations at all close to those we have been considering, the lethal dose can be calculated for periods of time other than one minute as a simple proportion. Fifteen milligrams per cubic meter of Tabun breathed for 10 minutes will do about the same job as 150 milligrams per cubic meter breathed for one minute.[69] If we now assume that the average Londoner would have a 10-minute exposure to Tabun delivered by a V-2 that landed nearby, we have to revise our estimate and suggest that the entire fleet of V-2s could have produced a concentration lethal to 50% of the people in 69% of London's area and severely damaging to most of those in the target area who managed to survive.

What countermeasures could the British have taken? Even if the Germans had saved their V-2 stocks for a massive late-January bombardment, several days would have been required for the total barrage because of the limited number of trained launch crews and launch equipment. This would have given the British time to evacuate the city.[70] In response to the V-1 bombardment, in fact, 1,450,000 people were removed to the countryside.[71] In addition, gas masks that were effective against Tabun had been developed; they would certainly have reduced casualties, if used in quantity.[72]

A sneak attack using Tabun, with the intention of killing as many Britons as possible, would more properly have been carried out by conventional manned bombers, which could have carried greater amounts to more concentrated targets in a shorter time. The V-2 would have been a better delivery system for Tabun than the V-1, because there was a much higher launch failure rate for the buzz-bombs than the rockets, and clouds of accidently released Tabun might have made things very uncomfortable for German V-1 launch crews. Although the V-2 might have played some role in delivering nerve gas, the primary delivery system should have been aircraft over long distances and artillery over short distances.

I have found no evidence that anyone ever considered using the V-2 as a delivery system for nerve gas or other exotic weapons. Von Braun's team designed the rocket around a 1-ton explosive warhead only because it seemed the only way of gaining the military's financial support for rocket development. They did not invest their energies in searching out the many ways rockets can participate in killing people. But if we are interested in a scientific analysis of the development of rocket technology, we have to be aware of possibilities that were not pursued as well as those that were actually realized. The use of Tabun warheads would not really have changed the economic balance between the manned bomber and the V-2. The bomber could carry more gas over a longer distance than the V-2 just as it could with explosives. For modern nuclear weapons, the physical size of the warhead is no longer the major issue, and both bombers and rockets have effectively infinite range. Because long-range rockets have been developed and are now taken for granted, we can hardly realize how dubious a weapon the V-2 actually was. The Tabun-carrying V-2 could not have annihilated England and won the war for the Germans, but the massive use of nerve gas delivered by aircraft would certainly have been an effective threat for ending the war in a truce in 1944 and might even have completely turned the tide for the Germans.

In principle, normal technological change happens through the competition of different technologies and inventions in an open market. The best solutions for each practical problem will win out over the worst, and there is an overall pattern of formal rationality to the development of technology. The Nazis seem to have failed to use an effective solution—nerve gas—for their pressing military problems. The weapons existed; they were simply not used. Had Hitler won the war with Tabun, the world would have faced a frighteningly different future politically from that of the late-1940s and 1950s, and the history of technological development over this period might also have been very different. In a way, the abnormal thing happened—an effective invention was withheld from use long after that stage in the development of general technology in which it was likely to appear. The nonuse of nerve gas in war is an example of a kind of Reactionary Technological Revolution—an implicit decision against change and development. In this case I am sure we are all happy to be reactionaries.

No satisfactory reason has been found or convincingly argued for the failure of the Nazis to break the presumably weak international norms against the use of gas in warfare which were adopted after the unpleasant experience of World War I. We can suggest a few important factors, however.

Brian Ford believes Hitler held back because of "the technological stalemate that existed."[73] The Allies also had gas, and Germany was sure they would retaliate in kind if nerve gas were used against them. The German writer Lusar agrees, declaring that both sides were

> ...quite conscious of the fact that if an aggressor power resorted to these annihilating weapons, there would no longer be any salvation for mankind. Therefore all powers held back from employing them, and only two occasions of their use are known, which however caused no casualties and were the result of mistakes in the distribution of munitions.[74]

According to one report, in mid-1944 the British were in a position to render 2500 square miles of Germany lifeless, through the use of gas. This could have included the areas of Berlin, Hamburg, Cologne, Essen, Frankfurt, and Kassel.[75] About this time Hitler ordered a step-up in the production of gas masks.[76] This explanation is based on the German fears of retaliation, not on the actual capabilities of the Allies, and the Germans probably were further ahead than they thought. The Nazi restraint makes sense only so long as they had some confidence they could still win the war. If they felt they were facing certain death by Allied firing squads, they would have been well advised to consider extreme measures despite the dangers that might follow.

There are other examples of weapons that were withheld for fear the enemy would begin to use them. Early in the war the Americans devised a proximity fuse for antiaircraft shells that greatly increased their effectiveness. For a number of months the shells were fired only over water, so the Germans could not retrieve a dud shell and learn to make their own.[77] Hitch and McKean have observed that, however difficult it is to keep military secrets from spies, a weapon that is in actual use cannot be kept secret.[78] The Germans copied the American bazooka antitank rocket in World War II and produced them in large numbers.[79] They used their own arrow-shaped projectiles primarily on the Russian front, afraid that the British or Americans would copy them if used in the West.[80]

A particularly good and simple example is given by the withheld invention of radar chaff. Two methods for blinding the enemy's radar were available to both sides in World War II: jamming the radar beams with strong radio signals, a variety of what is today call "electronic countermeasures," and dropping shreds of metal foil, or "chaff," which gives false images on the radar screens making it impossible for unsophisticated equipment to distinguish aircraft. The first method was effectively used by the Germans in February 1942, when they wanted to move several large ships through the English Channel.[81] The competition for electronic dominance was vigorously fought by both sides. The English successfully jammed the German radio beacons, which were employed to guide fleets of bombers to their targets, only to find the Germans using ever more sophisticated types of beacon which required even more advanced countermeasures.[82] But the early radar sets built by the Germans and in general use were not up to the challenge of chaff.

Churchill says that he had been aware of the feasibility of radar chaff as early as 1937, and that techniques for using it in combat had been entirely perfected by 1942.[83] He saw it as an effective means for blinding the enemy's defenses.

> But the snag was obvious. The device was so simple and so effective that the enemy might copy it and use it against us. If he started to bomb us again as he had done in 1940, our own fighters would be equally baffled and our defense system equally frustrated.[84]

Churchill learned after the war that the Germans had developed chaff as well and held back for the same reasons.[85]

Radar chaff would favor the attacker. It would have aided the Germans if used in 1940 at the height of their blitz of London, would have favored neither side (but made the cost of the war to each greater) if used shortly after then, and definitely favored the British when their bombardment of Germany was at its height. Accordingly, the British withheld use until they were convinced their side would gain more than the enemy even if both sides used it. First tried over Hamburg on July 24, 1943, the simple device was a complete success, confusing the enemy defense and cutting bomber losses nearly in half. This raid was the beginning of the concentrated week of bombing that in great measure destroyed the city, causing the world's first manmade firestorm in which temperatures approached 1000°F and 50,000 people were killed.[86]

Gas could have been used to end the war in a draw, but the Nazis failed to seize this option. Despite the resistance of Nazi ideology to notions of moderation, what the Germans really needed from 1941 to 1945 was an immediate peace treaty followed by a generation in which they could consolidate their massive gains and prepare for World War III. Every authority on the state of German military technology in the war agrees that research and development were cut back in favor of production of current weapons because the official plans all called for a quick war. In every way the war used up more of Germany's resources than had been expected. Nerve gas could have been used selectively to force the Allies to accept the German territorial gains. Hitler, the masterful propagandist of terror, could have given the British the choice of a favorable peace treaty or death. Without British participation, Russia might have lost the war, and America surely would have tired of it.

Today we are quite familiar with the concept of the balance of terror, the nuclear stalemate that has rendered direct war between the superpowers unlikely and has forced them to invent more subtle ways of battling each other. How clear this idea should have been to Nazi planners is hard to decide.[87] In *The Strategy of Conflict,* Thomas C. Schelling notes the slow and recent development of the idea of deterrence[88] and suggests that the tacit agreement of both sides in World War II to refrain from the use of gas may have been facilitated by the intellectual simplicity of the two clear options: gas or no-gas. The use of nerve gas would have required subtle bargaining with the enemy or a complex and unsure strategy for unilateral use.[89] The German Army had long engaged in war gaming, the invention and playing-out of many strategies in mock warfare; indeed, it has been said that the art of Kriegspiel was "most highly developed in the nineteenth century by the German General Staff."[90] However, the conduct of the war was clearly decided by Hitler and his inner circle of Nazi Party chiefs, and their unsubtle minds and romantic habits may have handicapped their participation in the invention of the balance of terror.

One might argue that some German leaders were simply too humane and moral to resort to nerve gas, and perhaps the German arsenal was created only as a counter to expected Allied progress in this area. Although some Nazi leaders, such as ex-chemist Robert Ley, head of the German Labor Front, proposed the use of gas;[91] others, like munitions minister Albert Speer and Luftwaffe fieldmarshal Erhard Milch, opposed this on moral as well as strategic grounds. Although we often think of the Nazis as evil incarnate and recall all too vividly the concentration camps, we fail to notice the extent to which the Nazis attempted until the very last to spare the German people. We can only speculate as to the various motives, but much evidence suggests that many Nazis were particularly concerned with gaining high social status through the party. High status requires a society that supports status differences and is not the same thing as brute power; status is connected with respect. The Nazis behaved dishonorably to people who were not members of the German social system and exploited those at the bottom of it, but their hesitation in using gas may have been partly motivated by a desire to retain the honor given them within their own nation.

The failure to turn to a strategy based on gas warfare may be in great part the result of an accident of history—the personal history of

Adolf Hitler. He had what amounted to a phobia about poison gas. As a footsoldier in World War I,

> ...he had been exposed to a slight attack of mustard gas. He immediately believed that he was blinded and speechless. Although he spent several weeks in the hospital, neither his symptoms nor the development of the illness corresponded to those found in genuine gas cases. It has been definitely established that both the blindness and the mutism were of an hysterical nature. The physician who treated him at that time found his case so typical of hysterical symptoms in general that for years after the war he used it as an illustration in his courses given at a prominent German medical school.[92]

This seizure of battle hysteria also served as a Shamanistic conversion experience.[93] From this time on Hitler had his mission, and much of his charismatic power came from the transformation of his mental illness into the intense projection of a will to power and the claim of personal perfection. Psychoanalytically, Hitler was not suffering from a mere phobia concerning gas. The gas incident was at the very heart of his pathological adaptation. Before Hitler could come to rational terms with his feelings about gas, he would have had to give up many of his neurotic defenses and abandon his magical role as *Führer*.

If Hitler's own psychological complex was indeed the major barrier to the employment of nerve gas in World War II, we have another extremely important example of the impact of personality on history. An historical determinist might argue that a second world war of some kind was bound to erupt between the industrial nations, but he could not have predicted that the dictator of one of the major contenders could have decisively influenced events in just this way because of irrational features in his personality. Because the kinds of social systems that would dominate the world were at stake, Hitler's psychiatric condition was the source of an indeterminism of profound significance in world history.

The technological history of the Third Reich makes it clear that Hitler and other Nazi leaders, with few exceptions, were scientifically and technically incompetent and made decisions on the basis of irrational intuition or other emotional criteria. Several examples are given in these pages. The failure to use nerve gas might be seen as but one more instance of incompetence. One of the features of the model presented in Chapter 3 outlining the process by which the German branch of the Spaceflight Movement achieved its purposes was that the patron should be relatively incompetent in technical matters. For the model to work, men like Adolf Hitler have to be fooled and, therefore, must be gullible and incapable of judging the worth of various kinds of technical invention. This means that whenever the model is working well, it is likely to describe a number of interactions concerning several innovations going on separately around the patron. Thus, we might describe a *technologically revolutionary situation,* defined as the presence of a rich and incompetent patron beset with problems that might be solved through technological innovation. In such a situation some developments may prosper unnaturally, while others will languish. From the standpoint of Nazi interests, the V-2 got much more attention than it deserved, while nerve gas got less. While the two inventions could have been combined, their combination does not really change the status of either with respect to other technologies competing with it. The idea of a nerve-gas-laden V-2 was important for us to consider but was never and would not have been a key factor in the outcome of the war or in the history of either technology taken alone.

When the American Joint Chiefs of Staff first discussed intelligence reports about the V-2 in late-1943, the Air Force representatives suggested that such a weapon might achieve a stalemate for the Germans if it delivered a new kind of explosive (atomic), highly lethal poison gas or bacterial agents.[94] On January 4, 1944 Dr. Vannevar Bush informed a British representative of his opinion that the German V-weapons must be designed to carry bacterial agents, because the payloads were too small for any other purpose.[95] While the Americans forged far ahead in biological warfare developments in World War II[96] and may have continued to hold the lead in this field until the present,[97] the Germans never got beyond the stage of experimentation and were in no position to attempt germ warfare, let alone successfully carry out a decisive attack delivering biological agents by rocket.[98]

Alternatives to the V-2

We have compared the V-2 with its most direct advanced-technology competition, the V-1, and now consider other *vergeltungswaffen* in comparison to the Peenemünde rocket.

The weapon which would have received the designation V-3, had it been successful, was the *Hochdruckpumpe* long-range gun, which Karl-Heinz Ludwig has called *ein Beispiel technischer Fehleinschätzung,* that is, "an example of technical misassessment."[99] In World War I, the Germans had received some success and considerable fame with their Paris Gun which had actually shelled Paris in 1918 from a distance of 130 kilometers.[100] Faced with the loss of air superiority in the last years of World War II, consideration of still larger guns that might have sufficient range to fire from the French coast to London was a quite natural step. Indeed, as early as 1928 Hitler himself considered this possibility. In a book written at this time but not published until after his death, Hitler looked forward to the next European war and imagined it might be between the English and French empires, which he saw as natural enemies. He stressed that the geographic location of France presented a tremendous threat to England:

> ...not only in that a great part of English living centers are so near as to be defenseless against French air attacks, but also because a number of English cities can be reached from the French coast by long range artillery. Yes, if modern technology is successful in producing a significant increase in the firing capabilities of the heaviest long range artillery, then even an artillery bombardment of London from French territory is not beyond the realm of possibility.[101]

Even before the war, the Röchling Iron and Steel Works, under the direction of an engineer named Gessner, had developed arrow-shaped projectiles for conventional guns. Because of their slim shape, these arrow-projectiles experienced considerably less wind resistance and therefore achieved a greater range than conventional shells.[102] In 1942 Chief Engineer August Coenders of the Röchling firm made a rough design for a special gun that might be able to fire his company's arrow-projectiles to a distance of 160 kilometers. Hermann Röchling himself became interested in the idea. What better way to ensure the sale of his company's products than to convince the government to invest in a wonder-weapon which depended upon them! Early in 1943 Röchling and Albert Speer discussed Coenders' project with Hitler, and in August Speer urged Hitler to order the construction of a special underground facility to fire it at the English, without even waiting for the results of test firings.[103]

The name given the new gun, *Hochdruckpumpe,* was half a cover designation and half pure description; it would indeed be a kind of high-pressure-pump. Boasting a 130-meter-long barrel, the *Hochdruckpumpe* would be built into the earth, permanently aimed at London. The arrow-projectile would be shot through the long barrel, impelled first by the usual charge at the rear. As it passed a number of short branch pipes on either side in the barrel, other charges would fire at precisely calculated intervals, pumping more pressure into the tube, each boosting the speed, until a velocity of 1100 meters per second was achieved. The projectile itself would be 2 to 3 meters long, weigh 140 kilograms, and carry 25 kilograms of explosives. The Germans hoped that such a gun could be fired every five minutes, and plans were laid for a massive battery of as many as 50 *Hochdruckpumpen* capable of showering the British capital with 600 shots per hour.[104] The battery would require a crew of 1000 men, and construction was immediately begun under a French hill.[105]

In January 1944 test firings of a prototype gave promising results, and Hitler ordered production of 10,000 projectiles a month.[106] He also decided that their warheads should carry incendiaries.[107] The *Hochdruckpumpe* began displaying flaws, and except for a successful but useless attempt to fire at Luxemburg when the Allies had occupied the area at the end of 1944, it gave no aid whatever to the Germans.

There were several problems: (1) The Allies forced the abandonment of the fixed battery pointed at London before it was ready for business. (2) The muzzle velocity was found to be too low to permit the projectiles to reach London. (3) The barrel had been designed with walls that were not thick enough to contain the shock of firing, and the *Hochdruckpumpe* frequently exploded. (4) The mass-produced arrow-projectiles were poorly designed and did not remain stable in flight.

A Professor Osenberg was called in to redesign the gun. He wrote Martin Bormann that the device was a general failure and blamed Coenders' "pure empirical testing-around without any scientific method."[108] At the beginning of the year Coenders had received the Fritz Todt Prize for special contributions to munitions development but by mid-year had been effectively discredited.

Almost every account says the *Hochdruckpumpe* was by nature an ineffective weapon.[109] Further, it might be vulnerable to air attack.[110] According to Irving's view, Coenders had exercised "an unhappy influence over the destiny of his promising invention, long after it should have become a purely service project."[111] Hitler had excluded the Army Weapons Office from the project, and Röchling's Coenders had been "reluctant to call in expert advice."[112] In 1945 a fact-finding mission of British Colonel T. R. B. Sanders concluded with the opinion that the *Hockdruckpumpe* could indeed have been a dangerous threat to London.[113]

In August 1944 Lieutenant Joseph Kennedy, older brother of John Kennedy whose later decision as President sent men to the moon, lost his life in an unsuccessful attempt to destroy the *Hochdruckpumpe.* He and another airman flew a stripped-down B-24 Liberator bomber stuffed with 10,000 kilograms of high explosive across the Channel with the mission of diving the plane into the *Hochdruckpumpe* emplacement which was so solidly protected by the hill under which it was built that the military thought only this extreme measure could do the job. Kennedy and his partner were supposed to bail out and let another plane guide the B-24 by remote control for the last few minutes. Before they were able to abandon the flying bomb, it exploded. One might speculate if John Kennedy thought back to his brother's death while considering investing in von Braun's team and the rockets developed from the V-3's spiritual brother, the V-2.[114]

The *Rheinbote,* or "Messenger from the Rhine," was a four-stage, long-range, solid-fuel bombardment rocket built by the Rheinmetall-Borsig Company as a direct competitor to the V-2.[115] Eleven and four-tenths meters long, it weighed 1715 kilograms on take-off but could deliver only 20 kilograms of explosives in its fourth stage to the target.[116] Unguided, it was stabilized by fins at the rear of each stage and was fired from a tracked arm on a motor vehicle. Its advantage over the V-2 was simplicity of operation, and for a solid-fuel rocket of that period, it was accurate and had the remarkable range of 160 kilometers.[117] Dornberger, unimpressed by the 1.2-meter hole in the ground produced by a test flight, commented: "And for *that* we burn 580 kilograms of gunpowder and throw two tons of iron into the neighborhood?"[118]

Dornberger urged that the *Rheinbote* be abandoned, but Hitler and Kammler, Germany's new Special Commissioner for long-range weapons, ordered it to be used in battle.[119] About 220 of them were fired at Antwerp—to little effect.[120]

The *Messerschmitt Me-262* jet fighter has been called "the most advanced production airplane" of its time[121] and "perhaps the only [aircraft type] that could have saved the Luftwaffe from defeat in 1945."[122] It had a top speed of 850-870 kilometers per hour and Allied air fleets were almost defenseless against it. In one battle of April 7, 1945, about 60 Me-262s shot down 25 Allied B-17s. In another attack at about the same time, a mere six Me-262s were able to down 14 B-17s without any damage to themselves.[124] There is general agreement that if the plane had been used earlier, even a few months earlier, the war would have been considerably delayed, perhaps until the Allies had their own jet fighter. Eugene Emme has suggested four causes that contributed to the unnecessarily late introduction of the Me-262:

> 1. Subordinate priority of the Luftwaffe in the German war effort during 1941, 1942, and 1943.
>
> 2. Problem of technical development and modification of the jet-type aircraft so that it would be operationally proficient.
>
> 3. The defensive nature of the German war-effort by 1943 which placed a host of pressing problems for decision upon the German High Command, and which magnified personal animosities in the Luftwaffe Command and the unhealthy competition among leaders of the German air industry.
>
> 4. The personal decision by Adolf Hitler in May 1944 to employ the Me-262 as a fighter-bomber (or "Blitzbomber" as it was called) which delayed its use as a fighter interceptor for at least six months.[125]

I am not convinced that the second cause was important, and I would correct the third to read:

> 3. Failure of the German leaders at any time to plan for a long war with the consequent policy that no new weapons were to be energetically developed unless they could be introduced into combat in a very short time.

Of course the V-2 was developed in some violation of this principle, as were other weapons. Furthermore, we would have to consider as valid reasons for German failure to develop some of the advanced weapons available to them:

> 5. The incompetent, amateur, romantic, antiscientific nature of the Nazi leaders and the culture they brought with them into dominance over German life.
>
> 6. The inward-looking, feudalistic, fratricidal nature of the Nazi social system, in which a new technical development, or other resource, was seen merely as another piece on the board of the game the Nazis were continually playing against each other.

The Messerschmitt corporation was given a contract to develop a twin-jet fighter as early as 1938,[126] but the world's first jet plane to fly was the Heinkel He-178, which took to the air in August 1939.[127] The Heinkel twin-jet fighter He-280 made its first flight in April 1940,[128] while the competing Me-262 did not get off the ground until July 18, 1942. Heinkel had trouble with his engines[129] and was out of favor with the Nazis;[130] so Messerschmitt's aircraft became the standard.

Generalleutnant Adolf Galland, one of Germany's most able and dashing air leaders, was attracted to the Me-262 by Willy Messerschmitt himself. Even the best Luftwaffe leaders, with few exceptions, made their decisions on the basis of *personal* rather than *technical* evidence, either through test flights they conducted themselves or by watching acrobatic demonstrations. Galland flew the Me-262 and was greatly impressed. He then attempted to interest Fieldmarshal Erhard Milch and Reichmarshal Hermann Göring in the fighter. Relations between Göring and Hitler were poor at this time, because the Luftwaffe had failed to win the Battle of Britain and could no longer even protect the fatherland, despite Göring's promise that his planes would completely shield German cities from air attack. Hitler was sinking ever further into his maniacal view of the war and was just turning to the V-weapons as means for outflanking and exacting vengeance against the English. Göring had constantly promised more than his air force could deliver, and Hitler was not interested in hearing more lies about an iron defense of the homeland.

Messerschmitt put on a demonstration of the Me-262 for Göring, who asked about the possibility of using the fighter as a bomber.[131] Göring was innocent of any understanding of technical matters; immediately after seeing the V-2 engine test he enthusiastically imagined trains and even ships propelled by rockets![132] In his continuing psychological depression and feeling that he had continually disappointed the *Führer,* Göring was impelled to see the Me-262 as a bomber, as a wonder-weapon that could satisfy Hitler's thirst for retaliation and as a means for restoring himself to good graces. Therefore, he went to Hitler with an enthusiastic tale about the possibility of using the Me-262 fighter to carry bombs.

In November 1943 Hitler watched a demonstration flight of the Me-262 in the company of Willy Messerschmitt and asked if the aircraft could carry bombs.[133] Messerschmitt casually replied that it could. Of course any aircraft can carry some kind of bomb, and Messerschmitt had made provision in his original designs for the Me-262 to carry one 500 kilogram bomb or two 250-kilogram bombs externally.[134] This does not mean the plane would have been an *efficient* bomber. Messerschmitt told Hitler the plane might be able to carry as much as 1000 kilograms of bombs, but that its speed would be cut by nearly 200 kilometers per hour by the drag produced.[135] Hitler is reported to have made a triumphant speech to the assembled air leaders, who included Göring, saying:

> For years I have demanded from the Luftwaffe a fast bomber which can reach its target in spite of enemy fighter defense. In this aircraft you present to me as a fighter plane I see the Blitz Bomber, with which I will repel the invasion in its first and weakest phase. Regardless of the enemy air umbrella, it will strike the recently landed mass of material and troops, creating panic, death and destruction. At last this is Blitz Bomber! Of course, none of you thought of that![135]

As Hirsch reports:

> The employment of the Me-262 as a fighter bomber was not decided on the basis of any tests, or any technical or operational basis, but solely on the basis of a bee in Hitler's bonnet that the Me-262 would become a bomber and nothing else.[137]

Hitler ordered that most Me-262s under construction be converted to bombers, but in secret opposition to his commands, Speer, Milch, and Galland attempted to continue construction of the planes as fighters.[138] When Hitler learned of this, both Galland and Milch were fired.

In the fighter configuration the Me-262 had a range of about 1050 kilometers;[139] comparable to the range of the standard Me-109, which was inadequate to provide thorough fighter escort for German bombers over London. The Germans had been slow to think of extending the range of the Me-109 with external fuel tanks, but surely would have been aware of the possibility of using this device on the Me-262[140] However, if the plane were already burdened with its maximum load in the form of external bombs, it could hardly have taken on the added weight and air resistance of external fuel tanks. If we assume that the bombs would have reduced range by 20%, the same factor by which they reduced speed, the Me-262 at most could carry a 1000-kilogram bomb load 400 kilometers and return. This actually compares favorably with the capability of the V-2, which could carry the same payload to a distance of about 250 kilometers.

Hitler's plan was to use the Me-262 for tactical bombing of troops, not strategic bombing of London, and considering the need for the plane as a fighter and the attrition England's air defenses would have caused on the slow in-coming flight, wasting the Me-262 as a substitute for the V-2 would have been foolish. Interestingly, even the apparently unsuited Me-262 jet fighter could have competed with the V-2 at its own game, and the fate of this fine aircraft was decided by the same irrational process of personal politics and intuitive thinking on the part of the chief Nazis as determined the investment in the V-2.

Mistel was the code name of an attempt to use conventional aircraft in a highly unconventional way. A reconditioned used Junkers 88 bomber, stripped of guns and crew facilities, would be combined with a single-seat fighter, either an Me-109 or FW-190.[141] The bomber would be filled with explosives and carry an auxiliary fuel tank bringing the range of the contraption to 2000 kilometers at a speed of 380 kilometers per hour.[142] The fighter would be set on struts piggy-back above the bomber, and the two would take off together. The pilot of the fighter would fly the connected planes until very near the target, usually a large fixed emplacement or a ship, and jettison the bomber, either on a collision course with the target or under radio guidance. The pilotless bomber would ram the target, and the fighter would race home to safety. The standard explosive was a 3,800-kilogram "shaped-charge" device which had a high penetrating power. Mistel was used with modest success against the British and American invasion fleet in 1944, and encouraged, the Germans prepared upwards of 200 Mistel bombers for use against the Russian electric power stations or the British fleet at Scapa Flow. These targets were never attacked, and despite some postwar interest in the idea, Mistel cannot be described as a great success. However, the scheme was not entirely crazy, and Mistel or the transformation of obsolete bombers into unmanned V-1 like flying bombs could be seen as real competition to the V-2.

The Germans were reluctant to adopt the *Kamikaze* tactic of their Japanese allies. According to one report, some German pilots were given the task of flying explosive-laden planes into enemy targets, but the pilot was always supposed to bail out at the last moment.[143] The Germans also experimented with *Rammjägern,* fighter planes designed to ram invading Allied bombers, and there was some thought of using manned V-1's for this purpose.[144] Wernher von Braun's long-time personal friend, Hanna Reitsch, one of the tiny handful of prominent women in the Third Reich, was one of the very few advocates of suicide interceptors. She backed up her opinion with personal daring and actually flew test flights in the V-1s herself.[145]

In our evaluation of the military value of the V-2, I believe we have shown that a number of other technical means existed that could achieve the same effect, some at much less cost. None of these extraordinary devices would have been needed by the Nazis if they had properly invested in conventional long-range bombers or if they had allowed themselves to make peace. From a military standpoint, the V-2 was a great mistake, however impressive its ultimate descendents, contemporary ICBMs, have become.

German Spaceships

If the V-2 was intended by its designers as a prototype spaceship, it had two deficiencies: It could not carry a man, and the velocity it could achieve was too low for real space missions. As early as 1940 the von Braun team had sketched plans for a successor rocket, the A-9/A-10, that would be a true space vehicle. In its early version this craft was to be a two-stage liquid-fuel rocket capable of attaining a velocity of 2800 meters per second and a range of 5000 kilometers. In its military version it would have been what we today call an Intermediate Range Ballistic Missile, or IRBM.[146] The A-9 upper stage was to be similar to the V-2, while the A-10 booster was a larger rocket of similar proportions.[147]

Von Braun asked Oberth at Peenemünde to perform a feasibility study and make initial designs for an *Atlantik-Rakete,* or "Atlantic Rocket," capable of bombarding the United States. In October 1941 Oberth made his report. He had calculated that a three-stage missile could be built with the necessary range but that the 1000-kilogram warhead, the same as that of the V-2 or A-9, would not cause enough damage to justify the expense of the missile. Such a bombardment would perhaps have a "moral effect" on the enemy, but "the expenditure was all out of proportion to the effect."[148]

Nonetheless, design studies continued. A second version of the A-9/A-10 emerged in which the upper stage had swept-back wings that would permit a glide of hundreds of miles after the A-9 had reentered the atmosphere.[149] To test the reentry and glide procedures, a few prototypes, essentially modified V-2s, were fired.[150] Because the German Army had not authorized the development of the A-9/A-10, von Braun justified the work by claiming that it was aimed at extending the range of the V-2 itself, and the test vehicles were given the cover designation A-4b, or V-2b. Extensive wind-tunnel tests were also performed on A-9 models with a variety of wing shapes.[151]

The late designs called for A-9 to have a pilot. Indeed, the vehicle could hardly have worked without one, considering the very primitive guidance devices the Germans had. A-9 had to reenter the atmosphere at a precise angle, then steer in a complex flight path to its target. The A-4b unmanned test rockets never achieved successful reentry. There was also a plan for the A-9 to have an auxiliary jet engine. By the time pilot, life-support, jet engine, and jet fuel are added to the vehicle, there is no room left for a warhead or bombs.[152] After the war von Braun was happy to assert the A-9 would never have been an effective weapon.

> An unbiased visitor to the planning group at Peenemünde would have heard little, if anything, discussed which related to other matters than reaching out into space....
>
> For the war-conscious officials, the object of the A-9 was explained as an extension of the range, to almost double that of the A-4. Calculations and computations as well as wind-tunnel data indicated that this might be achieved by utilizing the tremendous energy available after cut-off for a protracted aerodynamic glide. The wingless A-4 utilized this energy for destruction at impact. In point of fact, A-9 could extend its range only at the cost of velocity and would approach the target at subsonic speed. This sacrificed the proverbial non-interceptibility characteristic of A-4 and placed the A-9's tactical value in grave doubt.[153]

A topic of conversation that came up often in the von Braun circle in the late years of the war was the possibility of building a super-rocket, an A-9/A-10/A-11, in which the A-9/A-10 configuration would be boosted aloft by a still larger A-11. Von Braun and Ordway explain:

> The three-stage vehicle was to place the pilot of a modified A-9 into orbit. And there was even thought of an A-12 stage, producing a minimum of 2.5 million pounds of thrust; with the A-11 second stage and a winged A-10 third stage, it could possibly have orbited a payload of up to 60,000 pounds.[154]

This A-10/A-11/A-12 bears a resemblance to the ferry rockets sketched by von Braun in the spaceflight proposals he made to the American public in 1952.[155] It is also on the same general order as the actual Saturn I vehicle, designed by von Braun's team in Huntsville, Alabama, and first flown in 1961.[156]

Antiaircraft Missiles

The *Wasserfall* (Waterfall) was an antiaircraft missile derived from V-2 research and designed to be fired against Allied bomber fleets from fixed emplacements around German cities and other targets. By the end of World War II the rocket itself had reached an advanced stage of development and had been successfully test fired on several occasions.

Wasserfall stood just under 8 meters tall, had a maximum horizontal range of over 25 kilometers, and a top speed of 760 meters per second. After the war von Braun worked with an American team, developing a series of new versions of the Wasserfall under the name Hermes. Large American antiaircraft guided missiles, particularly the Nike series, are said to be direct descendants of the Wasserfall.[157]

Speer was enthusiastic about this missile, believing that the combined use of Wasserfall with jet interceptor aircraft could have broken the Allied air attack. According to him, Wasserfall was far enough along in 1942 that a crash program would have enabled them to introduce the missile into combat in great numbers before the end of the war.[158] The main obstacle in the way of its development was the confusingly wide range of competing antiaircraft guided missiles under development by various independent groups. Speer felt this diffusion of effort was a disastrous cause of waste and lost opportunities. Not until January 1945 was a competent man, Dornberger, appointed to head a high-level committee with the specific task of deciding the merits of the various schemes.[159] He decided in favor of the Wasserfall and the excellent, tiny, simple R4M unguided solid-fuel air-to-air missile, which we discuss below. He had a personal interest in both of these rockets.

Much of the design work on the Wasserfall was performed by von Braun's team at Peenemünde, and von Braun himself completed the basic design document November 2, 1942. The contract to design the airframe was given to a group at Karlshagen early in 1943, and because the only supersonic airframe design that had proven itself was that of the V-2, the Wasserfall was built with the same overall shape to save development time.[160] Göring himself may have placed the initial order for an antiaircraft rocket September 1, 1942, but it is not clear what events led up to this action.[161]

Let us perform a brief assessment of the Wasserfall. Although it could not be shot down in flight, the Allies could attack the vulnerable fixed launching sites with hit-and-run fighter aircraft. We might question the general worth of all early surface-to-air (SAM) missiles. The very latest Russian types employed in the recent Mideast War were highly effective in downing Israeli planes, but among their chief strong points were highly advanced guidance, immunity to electronic and other countermeasures, and mobile launching sites—all features the Wasserfall did not possess.[162] The other recent wartime test of SAMs was their use against American aircraft over North Vietnam; the rockets used were not so advanced or incorporated into as advanced a defense system as those employed in the Mideast conflict. According to the authoritative journal *Aviation Week,* the B-52 bomber losses averaged from 2 to 3% per mission, and an average of 60 SAM missiles were fired for every plane shot down.[163] Therefore, the Wasserfall could hardly have had much impact on Allied bomber fleets, even if it had been fully developed and deployed in great numbers.

A major failing of all the German SAM designs resulted from the fact that adequate guidance systems had not yet been developed, although engineers in various groups were working on a number of such devices. According to Pocock:

> Electronic computers were in their infancy in 1945, and even the optimistic Telefunken engineers did not envisage the construction of a computer which could establish the most economical interception course and control the radio command likely to hold the missile in this course.[164]

According to Theodor Netzer, one of the men working on the Wasserfall project, the war ended before any decision was made about what kind of device would be used to make the missile home in on its target in the last stage of its flight, and "the general control problem was far from being solved."[165]

The German plan for an SAM air defense system called for an initial stage in which 1500 missile batteries would be manned by a total of about 250,000 men.[166] Certainly, late in the war, Germany possessed neither the resources to build such a system nor the highly trained crews necessary to operate it. Although SAMs can, in principle, be effective when there are a few fixed targets to protect, such as ships, or in geographically limited theaters of war, they cannot compete with interceptor aircraft in range, effectiveness, or probably economy. The best current defense against attacking aircraft is probably interceptor planes equipped with air-to-air missiles, which the Germans almost had.

In one crucial sense the Wasserfall served as the test vehicle that permitted the development of a necessary part of space technology. Both the V-2 and Wasserfall were liquid-fuel rockets, but the chemicals used as propellants were quite different. The V-2's oxydizer, liquid oxygen, at ordinary pressures boils at 183° below 0 Centigrade, a difficult temperature to maintain in the rocket's tanks for any length of time. High-velocity American space booster stages, such as the Centaur, Saturn SIVB, or Space Shuttle orbiter, use liquid hydrogen as the fuel, cooled to below –217°C!

Such extremely cold liquids must be pumped into the vehicle shortly before launch; if there is a significant delay between fueling and firing, the liquified gasses boil away. Wasserfall used nitric acid as the oxydizer, a liquid that can be stored at room temperature. An added advantage of the Wasserfall's propellants is that they were *hypergolic*—that is, they ignited on contact in the engine rather than requiring a special igniting device as did the alcohol and liquid oxygen system in the V-2. The military advantage of the Wasserfall propellants is that the missile could stand ready to fire for days, waiting for enemy aircraft to fly within its range.

Most current American missiles use solid propellants. The American Nike-Ajax SAM, a child of the Wasserfall, which became operational in 1953, used liquid fuels in its main stage and had a solid-fuel booster. The Nike Hercules, which supplanted the less effective Ajax in 1958, used solid propellants in both stages. The experimental antimissile-missile Nike-Zeus, and its offspring Sprint and Spartan, use solid fuels exclusively.[167]

Long-distance space missions require rocket engines that can be used long after the launch, either to make course corrections or to permit landing on another planet. Propellants that facilitate the repeated use of the engine at precise moments without the unreliability introduced by complex ignition devices are additionally advantageous. Hypergolic, storable liquid fuels meet these requirements. In designing the Wasserfall, von Braun and his team were able to further spaceflight technology while appearing to be engaged in weapons research.

Wasserfall's oxydizer, red fuming nitric acid, was the propellant chosen by von Braun for his large booster rocket proposals, moon rocket, and Mars ship designs of the early 1950s.[168] Storable liquid propellants of a similar type were used in the Lunar Orbiter vehicles which mapped the moon, [169] and in the Mariner 9 probe which successfully charted Mars.[170] The Apollo Lunar Lander used nitrogen tetroxide instead of nitric acid, but the basic principle is the same, and the Wasserfall was clearly the first practical step toward the propellant systems used by all these space vehicles.

One can hardly think of a greater contrast than that between the 4¼-ton Wasserfall and the 4½-kilogram R4M air-to-air rocket, yet the latter was clearly the better weapon. General Simon writes: "[T]he outstanding armament achievement of Germany, the most important one that was developed to the practical working stage, was the R4M aircraft rocket."[171] According to Lusar, this unguided little rocket, fired from interceptor planes, was responsible for shooting down between 450 and 500 Allied aircraft, even though it was not introduced until early 1945.[172] This figure is not beyond question. The Germans habitually inflated reports of victories,[173] and historian Willy Ley has called Lusar "uncritical."[174] However, we are certain that the R4M was superb. The attacking planes could fire flocks of the little missiles from a distance of more than 1 kilometer, hardly risking the guns of the bombers.

The R4M showed an ideal balance of means and ends. It had been discovered that an explosive charge of only 400 grams was enough to destroy a B-17 bomber if it went off inside the aircraft, so the R4M was designed to carry 500 grams. The Wasserfall warhead was 150 kilograms in weight, 300 times as much.[175] This overkill was necessary because one could not count on a Wasserfall to come very close to its target, let alone be accurate enough to explode inside. Had Wasserfall been able actually to hit enemy bombers, considering its speed and mass, it would not have needed any warhead. In one series of accuracy trials, 50% of R4Ms were able to hit a 16-meter-square target at a range of 500 meters.[176] As many as 24 R4Ms could be mounted beneath the wings of an Me-262 jet interceptor, and with one firing system, 26 R4Ms could be launched from a FW-190 propellor fighter in the scant time of 0.3 seconds.[177]

General Simon suggests: "Many of the new weapons which the Germans placed on the battlefields of this war seem to have been motivated by...love of the spectacular,"[178] and notes that "instances were observed in Germany where great effort was devoted to a refined technique when the end result could have been accomplished by much simpler means."[179] The R4M was neither spectacular nor refined; the initial designs became a devastating weapon, "after an astonishingly short development time."[180] Indeed, although antiballoon rockets had been used effectively in World War I, the idea was ignored until a flight major from the front suggested such a weapon to Dornberger, who was able to assist its development.[181]

Unlike the Wasserfall batteries, the R4M was not a complete weapons system. It depended upon a delivery vehicle, in actual combat upon fighter aircraft requiring fuel, pilots, airfields, and a radar or spotter warning system. While Germany was running out of these assets, it at least had some, but the necessary guidance system for the Wasserfall remained a hope for the future.

One daring solution was to seek a compromise between the large antiaircraft SAM guided rocket and the conventional, rocket-armed manned jet fighter. The Bachem *Natter* (Viper) interceptor, a rocket with a man in it, was a crude vehicle designed to serve as a last-ditch defense against Allied bomber attack. On August 1, 1944 Knemayer, Chief of Development for the Air Ministry, was given the task of finding "a new means of approach. Was there something economical, fast, hard-hitting, and reliable that could be used to knock the enemy aircraft out of the sky?"[181]

Natter was designed by Erich Bachem, director of the Fieseler aircraft works, the company responsible for the airframe of the V-1.[183] Bachem was able to interest Galland and Speer in the project, but Göring stood in the way. The designer, "a man of great determination," then went to Himmler with his project.[184] Whether the SS leader actually saw military potential in the Natter, or merely used it as one more excuse for enlarging his power in Germany, Himmler backed the development of this device.

Like many other late Nazi designs, Natter was a response to scarcities, impotence, and a desperate situation. It also was an opportunity for the expansion of personal kingdoms within the feudalizing Nazi Empire. Bachem and Knemeyer were the *owners* of the *Bachemwerke* factory that was created apparently out of nothing to be the prime contractor for the Natter.[185] The craft was designed to operate without airfields, use substitute fuels, and be flown by minimally trained pilots. Even telephone poles could serve as launch towers!

> The scheme called for an extremely small airplane, to be made of the most readily available materials which, at this time, seemed to be wood and crude steel. The next requirement was simplicity of manufacture, so simple, in fact, that the planners wanted to press all cabinetmakers, furniture companies, and small metal shops into the construction program. Thousands of small manufacturers would mass produce the Natter in such huge quantities that, assuming a very low success in kill rate against the Allied force, by sheer numerical superiority the enemy bombers could be wiped from the skies.[186]

Although desperate, this approach would not have been ridiculous. Wooden construction was used on a number of successful aircraft in World War II. In the United States one of the best-known piano manufacturers turned to the construction of troop-carrying gliders built from wood.

The engine intended for use in the Natter was the inefficient liquid-fuel rocket motor developed by Professor Walter for the Me-163 rocket fighter, and the armament would be a cluster of unguided air-to-air rockets such as the R4M. The early test flights of the Natter were less than successful. One pilot reported that the craft flew well enough in the horizontal gliding test, but he had been forced to bail out, nonetheless. On March 1, 1945 Flight Lieutenant Sieber, a test pilot, became the first human being to be launched vertically in a rocket when he tested the Natter in its production version. He was killed.

Only 14 of the craft existed at the end of the war,[187] but 200 had been ordered from the Bachemwerke, 50 by the Luftwaffe, and 150 by Himmler's Waffen SS.[188] Natter was a crude anachronism: an attempt to blend highly advanced ideas with backward means and materials. Wasserfall was ahead of its time in the sense that practical development of such a complex SAM would be developed along the lines sketched for Wasserfall, but not for several years.

The V-2 after the War

The V-2 paved the way for postwar Russian and American developments. The early Soviet T-1[189] and American Redstone[190] missiles were

TABLE 4.2 THE V-2 AND ITS RUSSIAN AND AMERICAN DESCENDANTS

	German V-2	*Russian T-1*	*American Redstone*
Weight	13,000 kg	18,000 kg	18,000 kg
Engine thrust	25,000 kg	35,000 kg	34,000 kg
Height	14 m	19 m	19 m
Range	250 km	1000 km	320 km
Fuel	Alcohol	"Hydrocarbon"	Alcohol
Oxydizer	Liquid oxygen	Liquid oxygen	Liquid oxygen
Warhead	Conventional explosive	Atomic, etc.	Atomic, etc.

both developed directly from the V-2 and were remarkably similar to each other.[191] In Table 4.2 I compare the vital statistics for these three rockets.

Thus, although the V-2 was not effective as a weapon, its immediate descendants were, thanks to their atomic warheads. Similarly, while the V-2 was not powerful enough to boost a satellite toward orbit, the Redstone was and launched Explorer I, the first United States satellite in 1958.[192]

However, the V-2 did prove itself a highly effective tool for delivering scientific instruments to high altitudes on brief research missions. Beginning in 1949, the Soviet Academy of Sciences launched a number of Russian-built V-2 rockets in a program to explore the upper atmosphere.[193] In a highly publicized launch of May 1957, the Russians sent 2200 kilograms of instruments to an altitude of 212 kilometers. One particularly indistinct photograph of the carrier rocket in flight was widely distributed by Sovfoto and was subjected to speculative interpretations by Western writers.[194] The Russians are slowly publishing more and more information about the history of their space program, and a picture of the May 1957 rocket published in 1971 shows that it was clearly a V-2.[195]

In 1945 the United States collected a large number of German V-2s and rocket parts and fired 47 of them in an upper-air research program between 1946 and 1952.[196] Perhaps the most forward-looking part of the program was the Bumper Project in which a small

American-made liquid-fuel rocket, the WAC Corporal, was carried atop a V-2 as a second stage and ignited just after the V-2's fuel had run out. On February 24, 1949 a WAC Corporal carried 30 kilometers above White Sands Proving Ground in New Mexico, successfully fired its engine, and coasted up to a record altitude of 400 kilometers.[197]

Although the V-2 was inadequate as either a spaceship or a strategic missile, it served as the prototype for both and influenced the development of military technology in a direction that served the interests of the spaceflight enthusiasts and led ultimately to research and travel beyond the earth.

Chapter 5 / THE AMERICAN AND BRITISH INTERPLANETARY SOCIETIES

Of all the amateur rocket clubs founded in the 1930s, only two have survived, the American Interplanetary Society and the British Interplanetary Society. Science fiction historian Sam Moskowitz says they "were both launched by science-fiction editors, writers, and readers."[1] Harry Warner reports that science fiction clubs and amateur rocket societies frequently cooperated before and during World War II, but there was a parting of the ways thereafter.[2] Although the AIS and BIS began as thoroughly amateur clubs created by spaceflight enthusiasts and saturated with the fanciful dreams of science fiction, both clubs evolved into serious professional organizations in a matter of a few years. Today they are respected scientific and technical societies. As important parts of the Spaceflight Movement, they show the typical evolutionary pattern of successful social movements: born in obscurity as deviant outsider organizations, they grow and mature until they finally gain the status of conventional institutions—more or less parts of the Establishment.

The American Interplanetary Society

Amazing Stories, the first magazine devoted entirely to science fiction, was founded in 1926 by Hugo Gernsback. In 1929 he attempted to convince rocket pioneer Robert Goddard to become an associate editor of the magazine in exchange for stock. After giving the idea some consideration, Goddard rejected it.[3] Gernsback's motive was undoubtedly complex. He saw science fiction as a means for publicizing the findings and values of science; conversely, he could use real science in an attempt to publicize the fiction. Gernsback lost control of *Amazing Stories* but by heroic efforts was able to launch *Wonder Stories* to compete with it. He hired a former technical writer, David Lasser, to be editor of *Wonder Stories* and various offshoot publications.

On March 21, 1930, the day after his twenty-eighth birthday, Lasser called together 12 friends and founded the American Interplanetary Society. Lasser, like Gernsback, was sincerely interested in Spaceflight, but the founding of the AIS may in part have been an attempt to drum up publicity for *Wonder Stories.* The group that met to form the AIS included G. Edward Pendray, the historian of the Society and the only member to stay on into the space age, his wife, three young men who disappeared from sight shortly after, and seven fellows whom Pendray describes as established professional men:

> C. P. Mason, a writer and editor, who was the first secretary; Fletcher Pratt, the noted writer and authority on naval and military matters; Clyde J. Fitch, an engineer now [1945] connected with the international Business Machines Company; C. W. Van Devander, a newspaperman; Lawrence Manning, a writer and businessman,...Nathan Schachner, a lawyer and noted writer; and Dr. William Lemkin, chemist, teacher, and writer of textbooks in technical fields.[4]

As Pendray admits, this group could also be seen simply as Lasser's stable of writers, men who have gone on to achieve in various fields but were at that time united in their science fiction interests.[5] Day's *Index to the Science Fiction Magazines* gives us the following information:[6] In addition to being editor of *Wonder Stories,* beginning just at this time, Lasser was also editor of *Amazing Detective Tales, Air Wonder Stories, Science Wonder Stories,* and *Wonder Stories Quarterly.*[7]. C. P. Mason published four science fiction stories; the first, under the pseudonym Epaminondas T. Snooks, appeared in the May 1930 issue of *Wonder Stories,* the first edited by Lasser under this title.[8] Fletcher Pratt was a prolific science fiction author who first published in the May 1928 issue of Gernsback's *Amazing Stories.*[9] Lawrence Manning was the author of at least 14 such stories. His first was written with Pratt and appeared in the May 1930 *Wonder Stories.*[10] Nathan Schachner wrote dozens of science fiction stories; his first appeared in the Summer 1930 issue of *Wonder Stories Quarterly.*[11] William Lemkin was the author of at least 10

SF stories, first publishing in the November 1929 issue of *Amazing Stories*. Lemkin's third publication was a story in the September 1930 issue of *Wonder Stories Quarterly*.[12] G. Edward Pendray, writing as Gawain Edwards, published two science fiction novels and a novelette, beginning with the February 1930 issue of *Science Wonder Stories*.[13] I find no reference in Day's *Index* or in other, similar publications to Fitch or Van Devander.

Lasser attempted to convince the members of a New York science fiction club, The Scienceers, to merge their group with the AIS. Although its members were teen-aged boys, the Scienceers were in a sense the senior organization, because the club had been founded December 11, 1929.[14] None of the many articles on the history of the AIS published in its journals mention the flirtation with The Scienceers.

The first issue of *The Bulletin* of the AIS, dated June 1930, announced:

> Among the principal aims of the American Interplanetary Society are the promotion of interest in interplanetary exploration and travel, and the mutual enlightenment of its members concerning the problems involved....Since the creation of public interest is of prime importance, the Society has sought to awaken interest in itself as well as the ideas for which it stands.[15]

The Bulletin was edited first by Van Devander, then Fitch. Lasser became editor with the May 1932 issue, and the title was changed to *Astronautics*.[16] Lasser gave up this job about the same time he lost his job as editor of Gernsback's magazines.[17] In the meantime Lasser "wrote one of the first nonfiction works on rocketry, *The Conquest of Space*," published in 1931.[18]

The biggest early success of the AIS was a public meeting at which the eminent French rocket experimenter and aircraft manufacturer, Robert Esnault-Pelterie, was scheduled to speak. Two thousand people showed up, too many to be contained in the auditorium at the American Museum of Natural History that had been booked for the event. Esnault-Pelterie fell ill, and Pendray read the speech in his absence. Sections of the German movie *Frau im Mond*, based on Oberth's ideas, were shown as well; and to accommodate the crowd, the program was given twice. Despite the guest of honor's absence, the meeting was a thorough success and did much to publicize the AIS.[19]

In April 1931 Pendray and his bride visited Europe and traveled to Berlin to meet with members of the *Verein für Raumschiffahrt.* On April 12 he and Willy Ley, representing the *VfR,* signed a document of cooperation between the two spaceflight societies. There was talk of creating an International Interplanetary Commission.[20] Pendray inspected the German rockets, and despite considerable language difficulties, returned to the United States full of news and enthusiasm. On May 1 he spoke before a meeting of the AIS and reported on German liquid-fuel rocket designs.[21] Pendray later said this meeting "marked the beginning of liquid fuel experiments in this country, other than the work of Dr. Goddard...."[22] AIS member H. F. Pierce immediately proposed that the Society begin its own experimental program. The Society now had two competing foci of interest: the original philosophical romanticism and a new desire to begin practical work. The experimenters won out, and according to Milton Lehman:

> The society's space-fiction writers turned to other interests. One, the president of a nursery company, withdrew from extraterrestrial affairs. President David Lasser left space flight to become a union organizer and, in 1933, led a march of the unemployed in Albany, New York.[23]

In 1935 Pendray wrote an article for the journal of the British Interplanetary Society, urging his English comrades to follow the lead of the AIS and begin active experimentation. He admitted:

> When we actually began to consider experimenting, a number of our older members dropped out—thin-skinned creatures, evidently, who could not even stand the thought of putting our sacred declarations of faith to the test of the proving stand and the launching rack.[24]

After the first moderately successful rocket tests he could report:

> We have attracted many new members to replace the old ones who were frightened by the thought of experimentation. Our new members are men of importance and training. We now have an engineering and scientific society instead of a debating society.[25]

The AIS held biweekly meetings at New York's Museum of Natural History, and the gatherings included both technical discussions and inspirational exhortations. In February 1932 Pendray captured something of the idealistic excitement of that time in a statement he made urging his fellows to join in achieving the conquest of space:

> We offer you tonight an opportunity to take part in this tremendous development, a development beside which the most adventurous dreams of other ages are inconsequential. Men of the future will look back upon the era and will see little to interest them in our economic disturbances, in our small personal ambitions and accomplishments. But if the work of developing the rocket is successful they will view this decade as one in which a world revolution of transcendental importance began.[26]

The growth of the Society is no mystery. Individual young men were turning to interest in rockets and related fields. Dispersed as they were, joining the only existing organization was far more efficient than starting a new one. Most were at very early stages of their careers, and both their style and their hopes matched those of the members of the youthful rocket club. When the AIS was renamed the American *Rocket* Society in 1934, the editor of its journal explained: "In the opinion of many members, adoption of the more conservative name, while in no way implying that we have abandoned the interplanetary idea, would attract able members repelled by the present name."[27]

Pendray reports, however, that the *literary* spaceflight concerns of the old AIS faded and were replaced by more down-to-earth technical questions about rockets. "Young men with specialized engineering and mechanical background began to dominate the scene, while many of the original members dropped out."[28]

The first liquid-fuel rocket of the AIS was based on German designs. Like Goddard's first successful vehicle of 1926, the German *VfR's Mirak* and *Repulsor* rockets were built according to the mistaken notion that they would be more stable in flight if the motor were at the top and fired its jet downward, past the fuel tanks, pulling them upward.[27]

Mr. and Mrs. Pendray and H. F. Pierce were extremely ingenious in producing a test rocket with all the fixings for very little money. The 7-foot rocket itself cost all of $30.60, while the firing apparatus and accessories added another $18.80 to make a grand total of only $49.40! They showed great resourcefullness in adapting common articles as pieces of the rocket. Mrs. Pendray made a parachute of department store silk to protect the rocket on its return to earth, and a container for the parachute was made out of a 10¢ saucepan. The fuel was poured into its tank with a coffee pot.[30] The engine was to be water-cooled.

> The fabrication of the waterjacket gave us some difficulty until we discovered that the motor would fit nicely into a one-quart aluminum cocktail shaker which was just then being given away as advertising by a chocolate-drink concern.[31]

The rocket was taken to a farm near Stockton, New Jersey, and tested November 13, 1932, shortly after dark. The engine was fired for about 20 seconds while the vehicle was tied down, and developed about 60 pounds of thrust, burning gasoline and liquid oxygen.

> Suddenly we knew that the oxygen supply had been exhausted. There was an excess of gasoline, as we had planned. This now came spurting out, throwing a shower of fire all around the foot of the rocket and proving stand.
>
> We hurried out at once, fearing that the stand would burn up. Fortunately it had been thoroughly rain-soaked. When the gasoline had been consumed the fire was put out without trouble.[32]

Examination with a flashlight showed no damage to the rocket, but the weather turned from bad to worse, and after the rocket was accidentally dropped and bent, the tiny crew of experimenters decided to return home and build a new model.

The second rocket used a number of parts salvaged from the first and was fired from a beach on Staten Island, New York, on May 14, 1933. Several spectators were present, including representatives of the New York Fire Department and two news cameramen, who were told to stand 800 feet away. The stalwart members of the Society huddled in a trench 75 feet from the launch rack. The three heroic men who were actually to launch the dangerous vehicle were in a little dugout only 25 feet from the rocket. One was the "lighter" who was to ignite the engine by lighting a wick under the nozzle of the motor. A "valveman" was to open the valves that let the propellants pour in, by pulling a cord. A third man, a "helper," was supposed to assist the other two.

> Everything seemed to be going off well until Mr. Smith attempted to pull the cord which opened the rocket's valves. At this point the detachable lever, probably loosened by the wind, fell off the rocket. Mr. Smith

> courageously ran up to the rocket, where the fuse was burning merrily close to an oxygen tank that must by that time have been under pressure of nearly 300 pounds, and replaced the lever.
>
> In his excitement, or perhaps because of necessity, he then opened the valves before he had regained the shelter of his dugout. The photographs show that he was fully exposed at the time the rocket started. Fortunately there was no explosion at that point, and he was unharmed.[33]

The oxygen tank did explode, but only after the rocket had reached an altitude of 250 feet. It landed in the water 400 feet offshore and was recovered. A very amusing photograph shows four members inspecting the shattered debris; Mr. Smith has a particularly big smile.[34] Those were rough days. The Society was on the right track, but the first flight to the moon was a long way off.

One of the major problems of liquid-fuel rocket design is cooling; without some method for cooling the walls of the combustion chamber, it tends to melt, often at a crucial point, causing an explosion. In the April 1938 issue of *Astronautics,* member James Wyld described his idea for a regenerative motor, a kind of engine in which the fuel itself would be circulated around the combustion chamber before finding its way to the inside where it would meet and react with the oxydizer.[35] Both Goddard and the Germans had already developed rather large engines that used this principle, but they had kept their work secret, so Wyld was forced to invent it independently.

The first static test of Wyld's engine on December 10, 1938 aborted when there was trouble igniting the propellants. A second test later the same day produced a thrust of 90 pounds until the liquid oxygen ran out after 13.5 seconds. The firing crew was already adept at taking accurate readings of performance and reported on the Wyld results:

> These figures represent a great advance on those obtained in former tests, and are among the highest ever recorded. The fact that they were reached without severe damage to the motor is especially encouraging and definitely proves the feasibility of the regenerative method of cooling.[36]

These results stimulated great excitement within the Society and made Wyld something of a celebrity.[37] Another test conducted at Midvale, New Jersey, on June 8, 1941 was even more successful. Wyld's engine ran for 26 seconds without suffering any damage.[38] On August

1 the same engine was subjected to "exhaustive tests," fired three times for a total of 1½ minutes, producing up to 135 pounds of thrust.

In 1941 and 1942 "on the basis of this motor design"[39] and with some encouragement from the military, four members of the ARS Experimental Committee founded Reaction Motors, Inc., "the first company created solely to produce rockets."[40] Now a division of the Thiokol Chemical Company, Reaction Motors is a very successful enterprise, one of the major producers of rocket engines and related equipment in the country. In the beginning the company's officers were all taken from the ARS. The president of Reaction Motors, Lovell Lawrence, had held the job of club secretary; the secretary of the company, Wyld himself, was then president of the club. Wyld and John Shesta, who had been chairman of the club's experimental committee, constituted the engineering staff of Reaction Motors.[41]

The formation of Reaction Motors was another step toward professionalization and away from science fiction. As Pendray wrote, World War II "brought the society its greatest opportunity."[42] In addition to contracts secured by Reaction Motors, the ARS gained strength through the defense work performed by individual members in other companies.

Immediately after the war Reaction Motors built the liquid-fuel rocket engines for the Bell X-1, the first manned aircraft to fly faster than the speed of sound. A total of 6000 pounds of thrust was achieved by clustering four of the company's motors. In an article in which he reported on this design in 1947, Lovell Lawrence showed clearly that each of the four motors was a perfected version of Wyld's engine.[43]

In the December 1946 issue of its *Journal,* the ARS published a comprehensive index listing the 210 items published in issues 1 through 60 of the *Journal* and its predecessor, the *Bulletin.*[44] The items were listed by title and issue number, under 14 headings. The mean issue numbers of items under each heading gave a crude measure of shifts in interest of the Society. Table 5.1 presents the data for the eight categories that contained 10 or more items, presented in order from the earliest mean issue number to the latest.

The early mean issue figure for "interplanetary flight" reflects the early emphasis of the Society and the quick shift to more mundane or specific questions. The first 18 issues came out in rapid succession between June 1930 and April 1932 under the title *Bulletin of the American Interplanetary Society,* and according to a later editor, they "stressed

TABLE 5.1 ANALYSIS OF CONTENTS OF THE FIRST 60 ISSUES OF ARS JOURNALS

Category Given by ARS Editors	*Mean Issue Number*	*Number of Items*
Interplanetary flight	10.5	17
Theory and mathematics	34.1	17
Experimental reports, plans, news	36.0	62
History and biography	37.9	13
Motors and accessories	46.6	28
Jet power for aircraft	47.6	14
Bibliography and patents	51.4	10
Rocket weapons	52.7	15

the interplanetary aspect and noted many details dealing with space flight."[45] In 1944 the editor of the journal noted: "It has been some time since [we] have published any writing on interplanetary travel problems."[46] A year earlier the Society's president introduced an article on Max Valier's experiments by decrying the Society's early emphasis and praising its more recent turn toward limited practical questions:

> It is an unfortunate fact that some of the most interesting European rocket experimental work took place when this Society was engrossed in interplanetary matters, and paid little heed to such mundane things as regeneration, Dewar flasks, valves and fuel tanks.[47]

The size of the society increased at a moderate pace in this period. In 1938, of a total of about 100 members, only one third constituted the "active" membership, while the others were "associate" members. By 1944 the ARS had grown to 314, including 50 "active" members, 220 "associate" members who were presumably mere subscribers to the *Journal,* and 44 "junior" members, apparently teen-aged enthusiasts.[48]

On December 1, 1945 the ARS became affiliated with The American Society of Mechanical Engineers. Among the advantages accruing to the ARS were cited "immediate added engineering prestige" and "the likelihood of rapidly increased membership." The rocket society remained completely distinct from the A.S.M.E., but rented offices from the more established group and for years arranged to hold its annual meetings in conjunction with those of the mechanical engineers. In the official resolution of affiliation, the aims of the ARS were described as consonant with those of the A.S.M.E. and were announced as:

> ...to further the development of rockets and jet propulsion by the dissemination of technical information through the medium of meetings and publications and by other suitable methods....[49]

No mention was made of either interplanetary flight or rocket experiments. All 82 active members were sent ballots for a referendum on the question of affiliation. Fifty-two ballots were returned, 49 in favor, only three against. The tie to the A.S.M.E. was described by Pendray as a "momentous forward step"; it certainly marked a rite of passage for the ARS on its way toward respectability. The two ARS officers who signed the affiliation document along with Pendray were executives of Reaction Motors, and the move may well have had informal advertising implications for their company.

The December 1946 issue of the *Journal* contained a roster of active members, along with their addresses and brief professional descriptions.[50] These data could be tabulated. Geographically, the ARS membership was centered around the New York area, even 15 years after its founding in that city, and had just begun to develop a group of members in California, a state containing many aviation companies and some other groups interested in rocket propulsion. Table 5.2 shows the distribution for the seven states with more than five members and the

TABLE 5.2 DECEMBER 1946 ARS MEMBERSHIP BY STATES

Area	*Number of Members*	*Percent*
New York	40	28.8%
New Jersey	25	18.0%
California	12	8.6%
Pennsylvania	9	6.5%
Ohio	7	5.0%
Connecticut	5	3.6%
Massachusetts	5	3.6%
Other states (17), Hawaii, D.C., GPO	31	22.3%
Canada	2	1.4%
England	3	2.2%
Total	139	100.0%

more dispersed residue, which may include members who moved away from the New York area.

Tabulation of the employers of the active members shows a great emphasis on the aviation field and supporting industries. Some of those who worked for general manufacturing companies seem to have been engaged in aviation-related work, such as tool making or subcontracting. Those employed by chemical industries seemed primarily concerned with aviation and rocket fuel. Although the very brief self-descriptions may have been biased for the ARS audience toward representing the members as part of what we today call the aerospace industries, the data certainly support the idea that the membership was indeed deeply embedded in them even by this early date.

TABLE 5.3 EMPLOYMENT OF 1946 ACTIVE MEMBERS OF ARS

	Number	*Percent*
Aviation manufacturers	47	33.8%
Other manufacturers	22	15.8%
Military and government	14	10.1%
Chemical industries	13	9.4%
Airlines	9	6.5%
Probably self-employed	7	5.0%
Students	7	5.0%
Research laboratories	4	2.9%
College teachers	2	1.4%
No data	14	10.1%
Total	139	100.0%

Although these men were probably young and at early stages of their careers, a number would disappear from the membership rolls of the ARS over the following years. The successor to the ARS, the AIAA, published a membership roster in 1973, and in the intervening 27 years at least 84% of the ARS active members died or dropped out. I arrived at this figure simply by looking up each of the 139 names in the AIAA roster. Of course some of the 22 (16%) of the 139 names I did find in the later list of more than 22,000 may have been accidental matches, not really referring to the same people.

The most heavily represented single company in the 1946 list was

Reaction Motors, the ARS offspring, with eight employees. This is a mere 5.8%, but the influence was undoubtedly disproportionate; at the time Lovell Lawrence, Jr. was president of *both* Reaction Motors and the ARS.

By the beginning of World War II members of the ARS were showing interest in the new developments in air-breathing jet planes, a more conservative topic than interplanetary flight. The cover of the May 1942 issue of *Astronautics* shows a photograph of what was then falsely believed to be the world's first true jet plane, the Italian Caproni-Campini CC2. A number of the members worked on the development of American jet engines and aircraft during the war, and this new branch of aviation was a much more promising channel for personal career striving than was astronautics.

In 1947 the old charter of the ARS was amended to give great emphasis to "jet propulsion," by which was meant not only rockets, but even more importantly jet planes, their engines, and many aspects of their design.[51] The original prime focus, interplanetary travel, was erased. An editorial in the *ARS Journal* for July–August 1952 announced that the periodical was expanding its field of interest to include all aspects of jet propulsion.[52] Finally, the shift from rocket society to aeronautical society seemed complete after the January–February 1954 issue of the *Journal,* when it changed its name to *Jet Propulsion*.[53] The word *astronautics,* which had been the name of the journal on and off had disappeared from the masthead altogether; it would return again in the 1960s.

The 1950s were marked by rapid development of liquid-fuel long-range missiles, and of course the first *Sputnik* in 1957. The ARS grew explosively, from a membership of 1200 in 1951 to 12,000 in 1959. A year after Sputnik I the leadership decided to stress the rocket and space orientation of the Society again, and *Jet Propulsion* became the *ARS Journal*.[54]

In the early 1950s the ARS played an important role in promoting the earth satellite project—according to Bergaust and Beller, the decisive role.[55]

In 1953, noting that "a real interest in the subject of space flight ... has always persisted among many members of the Society," the Ad Hoc Space Flight Committee of the ARS made a report in the *Journal* urging a conservative but positive approach to the development of space travel. The committee took great pains to stress that spaceflight was a difficult objective and, almost pessimistically, commented: "It is believed, however, that space flight of inhabited vehicles may be achieved during this century if mankind is in a position diligently to undertake the task." The committee urged the ARS to "take a public stand in favor of realistic research and development looking toward the achievement of space flight," but stressed that "the cost and difficulty inherent in the achievement of spaceflight must never be underplayed."[56]

On November 24, 1954 the ARS submitted to the National Science Foundation, a brief but carefully written proposal urging the development of an artificial unmanned earth satellite. The proposal consisted of a number of brief essays by different authors outlining different advantages of the idea; it did not constitute a technical presentation.[57] The concept was hardly original with the ARS, and other groups were pushing such a project. The Vanguard satellites—tiny, light, basketball-sized minimum satellites—were the outcome, and nothing more grand was imagined.

The conservatism of the now well-established American Rocket Society did not please all the members, and some complained that space travel was being sacrificed to the mundane pursuit of rocket and missile technology.[58] "On November 20th, 1953, a number of professional and lay people interested in or participating professionally in fields related to astronautics..."[59] met to discuss the possibility of founding a counterorganization dedicated solely to space travel. Two months later The American Astronautical Society was founded, open to people "from all walks of life" and intending "to publicly disseminate and support as many of the creditable proposals for the conquest of space as possible."[60] Almost as an act of defiance, the journal of the AAS was named *Astronautics,* a title that had been used, then discarded, by the ARS

By mid-1955 the AAS counted 250 members and had begun to evolve from a mixed organization only partly professional to one much more like the ARS. Amendments to the group's constitution made "the membership requirements directly compatible with those of the other established scientific and technical societies in the United States."[61] The journal evolved from a popular-style news and propaganda magazine into a truly technical publication. The ARS remained the more influential group, however.

In 1961 the ARS held a highly publicized symposium and published the resulting book *American Rocket Society—Space Flight Report to the Nation.*[62] In undertaking to explain space to the public, the ARS took its place as the central organization in the field.

In 1952 the ARS had held a joint technical symposium with the Institute of the Aeronautical Sciences (IAS).[63] This organization was founded in 1932 by a group at MIT who "sent out a call to some 600 of the aeronautical profession, and 408 men" immediately joined.[64] Created in imitation of the Royal Aeronautical Society, the IAS was always a respectable organization of professionals. Among the early presidents were aircraft manufacturers Donald Douglas and Glenn Martin. The membership climbed in an almost straight line to a peak of 15,424 at the end of 1962. By this time the IAS was renamed the Institute for the Aerospace Sciences, a sign of its efforts to keep up with the times. Growth of the ARS, however, approached an exponential curve; starting from a smaller and less-professional base, the rocket society had reached half the size of the IAS by 1956 and surpassed it in 1959. Having had a reasonably happy courtship for 10 years, the ARS and IAS merged at the beginning of 1963 to form the American Institute of Aeronautics and Astronautics (AIAA). According to *Aviation Week* Editor Robert Hotz:

> [T]he Rocket Society was always livelier and more public-relations conscious than the staid high priests of the IAS representing a more established and then more prosperous segment of technology. Thus it was not surprising that in the merger the vitality of the Rocket Society staff...naturally dominated the new organization and the good grey men of the IAS staff faded gently into obscurity.[65]

In 1966, near the peak in government space expenditures and when AIAA membership was at its maximum of 33,000 members, AIAA official John Tormey sent a brief questionnaire to every member.[66] Nine thousand returned them, and Tormey was of the opinion that the sample was fairly representative; four of the questions are of interest to us. He asked respondents to indicate their ages and the highest academic degree they had received. For both of these questions 0.5% did not reply. Of those who did, 4.5% had no degree; 46.0% had received only a Bachelor's degree, usually, I believe, a B.S.; 36.0% had a Master's as well; a further 13.0% had earned Doctorates; 35.5% were 35 years old or younger; 53.0% were between 36 and 50; 11.0% were over 50.

When asked about their occupations, 4.7% said they were involved in teaching, 2.5% in university research, 7.5% in science, a whopping 61.0% in engineering, 23.0% in technical administration, and only 1.3% in completely nontechnical work. Respondents described the organizations they worked for: 7% colleges or universities, 69% industry, 5% military, 13% government, and 6% in the inevitable "other" category.

The AIAA has gone a long way since its birth as the American Interplanetary Society. No longer are the leaders science fiction authors and fantasts. By the mid-1940s they were aviation professionals; now the membership of the AIAA consists almost exclusively of aerospace engineers, executives, and others in established positions in the field. The leaders are no longer freethinking independent intellectuals; they are captains of industry or persons identified in one way or another with the interests of aerospace management. Consequently, the AIAA propagandizes for projects that can be achieved within a few years, using technology only slightly advanced over the current level, *projects that can be sold now*. The AIAA has become probably the foremost aerospace engineering organization in the United States and boosts aviation as well as conservative space projects.

To get a rough picture of the professional level of AIAA members and to contrast the rank-and-file with the leadership, I tabulated information from the AIAA membership roster for 1973, which includes a brief listing for each of its 22,500 adult members.[67] Of 117 committee chairmen for 1973, 115 were listed in the roster; for all 22,500 members the grade of membership was given, and in most cases there was a brief professional description. A comparison group, Members-I, was simply the 115 listings immediately above the committee chairmen in the alphabetical section of the roster. A second comparison group, Members-II is a random sample of 292 members listed in the roster.

There are four regular grades of adult membership in the AIAA: associate, member, associate fellow, and fellow. Appearing in the two highest categories, associate fellow and fellow, are 52.2% of committee chairmen but only 22.6% of Members-I and 24.4% of Members-II. I coded the 522 occupational descriptions according to whether the man appeared to be an academic, an employee, a member of management, or "unknown." It seemed to me that 48.7% of the chairmen but only 33.0% of Members-I and 33.2% of Members-II were in management positions. To correct for the possibility that my coding was biased, I produced a questionnaire including the 115 comittee chairmen listings and the 115 for Members-I and gave it to a panel of eight Harvard and Wellesley College students. They coded the listings into the same categories in a "blind" condition—that is, unaware what my hypotheses were or that there were two different groups in the listings. The average percent assigned to management from the 115 chairmen was 52%, while on the average, only 39% of the rank-and-file members were judged to be management.

There are two noteworthy points concerning these results: Committee chairmen are men of higher status in their profession than rank-and-file members and more likely to be representatives of management, and a surprisingly high percentage of members can be described as management. Of course neither I nor my panel of eight students had really good criteria on which to base our judgments, but the evidence is so striking it is hard to doubt these findings. Many of the AIAA members are described as department managers, vice presidents, or even presidents of corporations. Clearly the AIAA is a highly professionalized organization with close ties to management and most likely to represent the views and interests of management. It is not a trade union. It is not a group of deviant revolutionaries. It is no longer even an interplanetary society!

According to the AIAA membership application form, entry to Associate Fellow status is restricted to

> Persons who have been in charge of important engineering or scientific work or who have done original work of outstanding merit, or who have otherwise made out-standing contributions to the arts, sciences or technology or aeronautics, astronautics or hydronautics. They shall have twelve (12) years experience and be recommended by four (4) persons, three (3) of whom shall be Associate Fellow grade or higher.

Influence in the AIAA is given to those with high professional achievement, which in this day and age means also high bureaucratic position and clear identification with management. Influence is not achieved by social or ideological activism. A person who takes a strong stand on an issue may lose influence rather than gain it.

I have studied the biographical sketches of 16 nominees distributed by The New England Section for the 1973 and 1974 local chapter officer elections and found not a single clearly ideological comment, a stand on an issue—indeed nothing but laudatory summaries of the candidates' professional careers. Respondents to a 1973 AIAA questionnaire said the most important reason they belonged to the organization was to "enhance your professional career."[68]

In his study of the important German engineering organization, the *Verein Deutscher Ingenieure,* Gerd Hortleder argues that it is an ideological captive of industrial management, rather than the instrument of the general membership.[69] His complaint is that the *VDI* cannot act in a revolutionary social capacity or even play the role of a labor union for the general membership because it serves other interests. Naturally, we might wonder if such industrial captivity also makes such an organization less likely to play a *technologically revolutionary* role.

The *VDI* provides an excellent example of a counterrevolutionary stance toward brand new ideas. According to Willy Ley, a close associate of Oberth, when Hermann Oberth's rocket writings appeared on the German scene in the mid-1920s, they did not receive a warm welcome from established scientific organizations.[70] In May 1927 Dr. H. Lorenz, a leader of the *VDI,* published in its *Zeitschrift* a vehement attack on Oberth's ideas in the form of a spurious mathematical refutation of the idea that rockets could acieve spaceflight velocities.[71] He calculates that a space rocket using nitroglycerine as fuel, would have to be built with only 0.5% of its weight devoted to engine, crew, and superstructure, and 99.5% given over to fuel. If the most favorable fuels, hydrogen and oxygen , were used, these propellants would still have to make up 97.1% of the launch weight of the vehicle.[72] Impossible! Of course Dr. Lorenz had ignored the now commonplace idea that rockets could be built in several stages, a plan that Oberth had already shown could solve the problems to which Lorenz attached so much importance.

In a later note Dr. Lorenz mentions the idea, put forth by Oberth, of "rockets set one inside the other, that would be cast off one after another," and vaguely dismisses such a complexity as unworkable from an engineering standpoint.[73] Oberth and Dr. Hohmann, a member of both the German rocket society and the *VDI,* wrote replies proving Lorenz wrong, but both were rejected for publication by the *Zeitschrift.*

In this same period Lorenz, was publishing other articles on topics such as gas dynamics in the *Zeitschrift des VDI*. He was in the position of established expert on just those branches of physics closest to rocket engine technology. Oberth was able to interest a much younger German engineering organization, the *Wissenschaftliche Gesellschaft für Luftschiffahrt,* an innovative aviation group. At their 1928 meetings Oberth and Lorenz debated each other; Oberth came out best.[74]

Following Hortleder's analysis, I believe we can say that the current *VDI* is very much like the AIAA. Much of the financial support for both come directly from industrial companies, and both are the captives or servants of industry. The AIAA seems proud that only 17% of its 1972–1973 income of $4,200,000 was derived from membership fees and mentions that $190,000 came from corporate member dues, paid directly by companies.[75] In response to a recent AIAA questionnaire, members urged that corporate membership be continued; 51% took this position against 23% who wanted it eliminated. When asked their opinion of "the impact of corporate membership on AIAA's responsiveness to individual member needs," 25.1% had no opinion; 28.1% thought there was no significant effect; only 13.2% felt responsiveness was enhanced, while 32.1% felt it was impeded.[76]

The 1972 and 1973 published financial reports of the AIAA disclose its sources of income, which as Table 5.4 shows, were heavily weighted toward government and industry.[77] "Member services" include both personal and corporate dues, and "publications" and "meetings" are sources of revenue from both individuals and companies. The NASA contract is primarily for a technical information service, and the Department of Labor pays for the AIAA's only significant "social service," a program to retrain and find jobs for unemployed aerospace engineers. Of course the fact that the AIAA is greatly supported by industry and government agencies does not necessarily doom it to abject servitude to these sponsors, but taken with the other evidence presented here, it does support the opinion that the AIAA is no longer free to examine and propound revolutionary ideas. Recently the Eastern Sociological Society conducted a membership campaign which focused on this issue:

> In this age of self-determination, it is disquieting that the E.S.S. must raise a high proportion of its income from commercial sources. These funds support the Annual Meeting. A larger membership would enable the Society to be more independent. A society of scholars who are

TABLE 5.4 SOURCES OF THE AIAA'S FINANCIAL SUPPORT

Source	Fiscal Year 1972	Fiscal Year 1973
Member services	$1,013,200	$974,600
Publications	614,700	609,300
Meetings	523,900	670,800
Library services	116,000	114,900
Investment income	104,900	131,800
Public affairs	—	1,000
Washington office	—	15,000
NASA	1,429,100	1,371,100
Department of Labor	491,200	396,000
Total	$4,293,000	$4,284,500

concerned about objectivity, honesty, and integrity ought to be self-sustaining, if possible, to prevent improper influences.[78]

That the AIAA has polled a sample of the membership frequently through questionnaires in recent years is interesting, because the Institute's election process absolutely does not test member sentiment about any issues. An election on issues or a direct referendum would be binding, while a questionnaire only advises. The leadership of the AIAA appears to be self-perpetuating. New candidates are nominated by old officeholders. There was even an embarrassed move recently to change the wording of the AIAA constitution because the article on election of officers "tends to mislead the reader to think he has more voice in the election of the Corporate Officers of the Institute than is factually the case."[79] Mind you, the amendment would not have given the common member more power, but merely informed him that he had little!

In response to AIAA questionnaire inquiry, only 32.6% of the membership polled felt they had an adequate "voice in overall AIAA affairs," and only 27.7% felt they had an adequate "voice in establishing slates for and election of national officers."[80] We might well imagine that these satisfied members are primarily the third of the 22,500 we might call management, but the AIAA data are not broken down in a way that allows us to test this hypothesis.

According to the bylaws of the AIAA, candidates for office in the national organization are nominated by a committee appointed by the current officers. An independent candidate can get on the ballot, if supported by a petition signed by 300 or more voting members, but clearly the system of power within the Institute is thoroughly self-perpetuating and based on professional seniority.[81]

The journals of the AIAA and the many technical meetings are concerned with scientific and engineering topics. When advocacy appears, it is almost invariably in the form of proposals for new aerospace projects in which the government is urged to invest immediately. Rarely is the consensus broken, and nothing approaching a disagreement is printed in the journals, unless it is a narrow debate over the best technical means for achieving some aerospace objective.

Earlier we tabulated the entries in the 1946 index of articles in the ARS journal, and were able to use the results as an indicator of shifts of interest over time. Reasoning that an analysis of more recent data would be worthwhile, I did a similar tabulation of the 1974 index to the publications of the AIAA.[82] It contains 2048 entries, but a single article is often listed more than once. Thus, it is more a measure of topics than articles per se. The index is broken up by major headings and subheadings under them. Some of the categories describe articles with applications for both aviation and space; for example, the largest category is *fluid dynamics* with 565 entries (27.6%), many of which refer to topics of equal importance for both halves of aerospace. Other headings apply only to aviation or only to spaceflight and can be used as a measure of the stress the AIAA puts on each. *Spacecraft technology* includes 300 items, 14.6% of the total; *aircraft technology* includes 304, which is 14.8% of the total. Here the balance appears almost equal. The overall heading *propulsion* includes subheadings which can be interpreted as follows: 35.5% of the Propulsion items concern aviation and marine topics; 34.5% could relate to either aviation or space questions; and 30.0% are clearly about spaceflight only.

The AIAA journals appear to have one editorial foot in aviation, the other in spaceflight. Another way of summarizing the data would be to say: The AIAA is firmly embedded in the Aerospace Complex. Dominated by aerospace management, it is a technical society overwhelmingly concerned with the down-to-earth, day-to-day competition for government contracts. No longer the American Interplanetary Society, it has participated in the maturation of the Spaceflight Movement by becoming thoroughly institutionalized and bound into the Establishment. Unlike its British cousin, the AIAA has lost much if not all of its revolutionary potential.

The British Interplanetary Society

Accounts of the history of British spaceflight development usually begin with the career of Sir William Congreve (1772–1828) who spent much of his life working on solid-fuel military bombardment rockets. Congreve's rocket artillery produced "the rockets' red glare" over the American Fort McHenry in 1814.[83] Although Congreve does seem to have been interested in the idea of space travel from the age of 13 until his death, there is no evidence that he saw rockets as the proper means of propulsion for spaceships.[84] Aware that the armies of India had been using rockets and inspired to action by the threat of an invasion by Napoleon in 1804, Congreve invested his own money in a study of already-existing British rockets. His father, Major-General William Congreve, was Controller of the Royal Arsenal at Wollwich, and with the support of the Ordnance Board, the younger Congreve was able to conduct a considerable series of experiments to improve the war rocket.[85] Willy Ley calls Congreve's influence "enormous" and suggests that his successes lead to the establishment of rocket batteries or even independent rocket corps in the armies of 15 European nations.[86]

A second important British rocket pioneer was William Hale (1797–1870) who improved on Congreve's developments. His most notable invention was spin-stabilization for rockets which previously had been guided only by the heavy sticks familiar to anyone who has launched a conventional fireworks rocket.[87] Spin-stabilization is very commonly used for unmanned earth satellites and was the only method of guidance employed on the third stage of the Vanguard and second through fourth stages of the Jupiter-C which launched the first American satellites. The first American rocket artillery was not of the Congreve type, but it was sold to the United States by Hale.[88] His weapons were used in the 1846–1848 Mexican War and the 1861–1865 Civil War. In a demonstration conducted in November 1862, a Hale rocket accidentally threatened the life of Lincoln himself.[89] Much of Goddard's early rocket experimentation was aimed at further development of conventional solid-fuel bombardment rockets, and many early spaceflight pioneers were informed and encouraged by the successes of Hale and his contemporaries. Although Hale is not known to have had any interest in spaceflight, his work did contribute to the general development of rocketry.

A number of factors have hampered British rocket development, not the least of which has been the difficulty of finding room for testing. William Hale was forced to move from place to place at least 10 times in search of "an extent of ground over which to fire, quite free from buildings and cattle."[90] The ancient Guy Fawkes Act[91] or the Explosives Act of 1875[92] put formidable legal barriers in the way of anyone hoping to fire a civilian rocket in Britain. In the 1930s P. E. Cleator and others in the British Interplanetary Society attempted to convince their government to allow the Society to engage in rocket testing on English soil; their appeal was rejected.[93] As recently as 1972 P. J. Parker complained that British law makes the launching of even the most tame American commercial toy solid fuel rockets almost impossible.[94] The British military did conduct tests of bombardment rockets in the home islands, but in 1939 Sir Alwyn Douglas Crow, Director of Ballistic research at Wollwich Royal Arsenal where Congreve had worked 130 years earlier, was forced to move his firing range to Jamaica.[95]

Throughout 1933 a 25-year-old resident of Liverpool, Philip Eliaby Cleator, struggled to create a British counterpart to the American Interplanetary Society. He succeeded in publishing an article on "The Possibilities of Interplanetary Travel" in the January issue of *Chamber's Journal*, but the public paid little attention. Other attempts at publicity failed. Finally, Cleator was able to convince the editor of the *Liverpool Echo* to print an appeal for members. The number of replies was underwhelming: *one.*

Moore Raymond, a correspondent for *The Daily Express,* grew enthusiastic about Cleator's project and was able to get a mention of the British Interplanetary Society onto the front page of his newspaper. On October 13 the BIS was officially founded with a total membership of six.[96]

Using his own money, Cleator published the first issue of the *Journal of the British Interplanetary Society,* dated January 1934. This slim pamphlet informed the reader about the new Society, mentioned a little news, carried a full-page ad for Chamber's publications, and bore on its cover a romantic drawing of a spaceship rising up between the spires of a futuristic city. The membership now totaled 15, if three honorary members are counted.

Like the American Interplanetary Society, the BIS contained a number of science fiction fans and authors in its earliest years. One of the 15 founders, Leslie J. Johnson, was an avid fan, and author Eric Frank Russell joined soon after.[97] Cleator himself published at least three short science fiction stories, beginning in 1934.[98] Like real space developments, science fiction magazines have not flourished in the constricted British market, and the BIS was therefore less firmly embedded in the young science fiction subculture. The founding group included two radio experimenters who operated transmitters, an automotive engineer, a young lady who served as secretary, a gentleman named Binns who was able to provide a meeting place, and "four young but most enthusiastic members," young fellows not yet entered upon careers.[99]

Cleator ran out of money and was unable to finance a second issue of the journal. Member C. H. L. Askham approached an old friend of his, millionaire John Moores, who agreed to finance the second and third issues under the condition that his name would not be used. Cleator says: "...Moores was more than willing to answer the call of his friend, but he wisely shrank from open association with a group of half-wits who were planning to lay hold of the Moon![100]

However one wanted to express it, Cleator agreed: "The ultimate aim of the Society, of course, is the conquest of space and thence interplanetary travel."[101] Looking backward, he and other members have been quite ready to admit that outsiders considered them crazy.[102] Val Cleaver, who joined the BIS in 1937 and went on to become a leading rocket engineer, has described the prewar Society as "a small group of enthusiasts and cranks." He adamantly declined to say which of the two nouns described him.[103] H. E. Ross said the behavior of the early BIS was "pixilated." "And why not?" he asked. "We were unorthodox."[104]

The handful of members found it convenient to meet "in the upstairs room of a cafe in Liverpool's Whitechapel, an unpretentious place, which exactly suited our needs: it was centrally located; it remained open until late hours; it provided light refreshments as and when required; and its charges were absurdly cheap."[105]

When the American Interplanetary Society changed its name and became a *rocket* society, the British group briefly considered a similar change, but rejected it. As Cleator argued: "The *raison d' être* of the Society—however remote it may seem at present—is to achieve the conquest of space, and thence interplanetary travel."[106]

At about the same time Cleator flew to visit the German rocket society at Berlin, and like Pendray of the American society, conferred at length with Willy Ley and was introduced to a number of other prominent men in the field. The second issue of the *Journal* carried news on some of the developments in the United States, Austria, France, and Germany. The Society has always had an international flavor and served as a clearing house for news and ideas from many countries. The third issue even carried a little news about Russian developments.

Near the beginning of 1935 G. Edward Pendray of the American Rocket Society suggested that the BIS join with his group and the Cleveland Rocket Society in Ohio to publish a joint monthly journal instead of the modest pamphlets each had been able to produce. Each group was to put in one third of the total expense, a contribution of something over 50 British pounds each. The idea was dropped because, as Cleator explained: "There can be no question but that it would be money well spent, but we, for one, cannot afford it...."[107]

Cleator published a popular book *Rockets Through Space,* which presented serious ideas about spaceflight in an optimistic and folksy style. It contains a number of nontechnical discussions of technical questions that must have been very educational for the average reader. For example, Cleator joined all the major pioneers in arguing for liquid fuel over solid fuel and noted that hydrogen and oxygen would be the most powerful conventional propellants.[108]

Rockets Through Space enjoyed considerable publicity and was given an added boost by the science fiction movie *Things to Come.* An advertisement for Cleator's book quotes from a review in the *Manchester Guardian* linking it with the film.[109] The BIS benefited from both the film and the book. Society member D. W. F. Mayer complained that the spaceship launched at the end of *Things to Come* was based on the wrong technology because the hardy astronauts were fired from a huge gun. The film was valuable as prospace propaganda if not as an engineering blueprint.[110]

Made in 1936, *Things to Come* looks forward to World War II, which begins as Britain is subjected to a sneak attack by bomber fleets. The cruel conflict grinds on and on, past 1945, to an inconclusive and disastrous end around 1966. Mankind is reduced to a barbarous existence in which political power is held by feudal chieftans. All nations have collapsed, and technology is a confused mixture of the Middle Ages and the twentieth century.

The story follows the members of a single family over the period of a century, but the real subject is the fall of one civilization and the rise of a new one. At the most dismal moment, a new fleet of advanced aircraft sweeps across Europe, bringing the strength of Reason from hidden Technocratic bases to install the Rule of Science and inaugurate an Age of Gold. The last scenes take place in 2036, when the barbarism of the twentieth century is almost forgotten and some romantics call for an end to Progress and a return to Nature.

The first spaceship becomes a political issue. The Technocrats are about to fire the first astronauts toward the moon, but the Romantics rush to the launch site to block that evil deed with their bodies. The film makes a strong statement against the Romantics in favor of science and technology. The spaceship is the symbol of human progress. The launch is made just in time, and the fathers of the two astronauts stand before a huge telescope and discuss the future of Man:

PASSWORTHY: Oh, God, is there ever to be any age of happiness? Is there never to be any rest?

CABAL: Rest enough for the individual man—too much, and too soon, and we call it Death. But for Man no rest and no ending. He must go on, conquest beyond conquest. First this little planet with its winds and ways, and then all the laws of mind and matter that restrain him. Then the planets about him, and at last out across immensity to the stars. And when he has conquered all the deeps of space and all the mysteries of time, still he will be beginning.

PASSWORTHY: But...we're such little creatures. Poor humanity's so fragile, so weak. Little ...little animals.

CABAL: Little animals. If we're no more than animals we must snatch each little scrap of happiness and live and suffer

> and pass, mattering no more than all the other animals do or have done. [He gestures grandly up at the stars.] Is it this—or that: all the universe or nothingness. Which shall it be, Passworthy? Which shall it be?[111]

The membership of the Society grew, stimulated by the film and Cleator's book. On October 27, 1936 the first meeting of a London branch was held. Arthur C. Clarke, later to become famous as an author of science fiction and popular books on science, was elected branch treasurer. Obviously inspired by the scene from *Things to Come* quoted above, Clarke wrote in one of the BIS leaflets:

> Looking out across immensity to the great suns and circling planets, to worlds of infinite mystery and promise, can you believe that Man is to spend all his days cooped and crawling on the surface of this tiny Earth—this moist pebble with its clinging film of air? Or do you, on the other hand, believe that his destiny is indeed among the stars, and that one day our descendants will bridge the seas of space?[112]

Only one third of a century after *Things to Come* and two thirds of a century before it had predicted, the first crew of astronauts reached the moon's surface. When Armstrong, Aldrin, and Collins returned to Earth, they received a congratulatory message from The British Interplanetary Society, which concluded: "When man has conquered all the depths of space and the mysteries of time, then will he be but still beginning."[113]

Cleator's influence faded. Some members resented the degree to which he dominated affairs in the beginning; but when others in Liverpool tried to take over his job, they balked at the tremendous dedication needed. By the beginning of 1938 the London branch had taken over publication of the journal and was named the official headquarters. Its membership had grown to twice that of the Liverpool branch, and of course London was the obvious site for a headquarters. Unlike its American counterpart, the BIS was never able to conduct practical experiments with rockets and was primarily a club of space enthusiasts whose only public function was propagandizing for spaceflight and spreading information.[114]

At its prewar height, the BIS had slightly more than 100 members and boasted an annual income of about $300. Arthur C. Clarke estimated that the "total amount spent on the British space effort before the outbreak of war was less than $1000."[115] Clarke describes the membership:

> There were about ten of us in the hard core of the society, and we met at least once a week in cafes, pubs, or each others' modest apartments. We were almost all in our twenties, and our occupations ranged from aeronautical engineer to civil servant, from university student to stock exchange clerk. Few of us had technical or scientific educations, but what we lacked in knowledge we made up in imagination and enthusiasm.[116]

The move to London was the opportunity for a renewal of the Society's momentum and marked a step in its maturity. Professor A. M. Low, a man of some professional stature who had long been interested in the problem of spaceflight, was elected president. In the first London issue of the Journal, R. A. Smith announced the new policy of the leadership:

> We have decided to concentrate our attention on the task of meriting a reputation for sound scientific work, by undertaking a survey of the whole problem [of spaceflight], such as will attract the interest of the scientific world and command the respect of the layman.[117]

The two issues of the *Journal* that came out in 1939 reported on an important design project completed by the nine-man technical committee: the famous first *BIS Spaceship.*[118] The committee had decided to sketch a spaceship capable of carrying two or three men to the moon and back, using solid rather than liquid fuels. This backward step can be explained in part by the fact that the members wanted to present a sober and therefore "conservative" plan to the world and in part because they had no experience with liquid-fuel engines themselves and had been somewhat discouraged by the reports from American experimenters. They knew little about Goddard's most recent work, were just about to learn of the breakthrough achieved by the American Rocket Society, and had not the slightest idea that von Braun's team was already well on the way to building the huge V-2 liquid-fuel rocket.

The BIS Spaceship was to stand 32 meters high with a diameter of 6 meters. The launch weight was estimated at 1000 tons. Divided into six stages, the ship was to contain 2490 separate solid-fuel rocket motors, each of which would drop off after it had fired. A complex switching system had been designed to fire the rockets and separate the stages in proper sequence. Many design details had been sketched out, including a system for steering the craft, spaceship cabin facilities, and procedures for landing on the moon. However, none of the descriptions of the vehicle indicate that thorough calculations had ever been made to determine if the five booster stages really could give the sixth stage enough velocity to complete its mission. The BIS Spaceship was a conceptual design, and the innovations were all in the area of mechanical details, not engine performance or trajectory calculation.[119]

Unable to begin construction of even a scale model, the technical committee contented itself with work on some of the navigation instruments that might be required for a lunar journey. There was some experimentation with primitive intertial guidance instruments similar to those later used in the V-2—gyroscopic devices that would measure acceleration and speed without the need to take a sight on anything outside the ship. The BIS actually built a remarkable if frivolous bit of machinery called a *coelostat.* This contraption was essentially a periscope with a rotating element that enabled a person to look at a rotating object and see it as if it were standing still. In demonstrations a rotating disk with the letters BIS could be seen as if stationary, by looking through the coelostat's eyepiece. The cabin of the BIS Spaceship was supposed to rotate to provide artificial gravity for its passengers, and the coelostat would be used as a periscope to allow the navigator to view the stars as if the ship were standing still.[120]

The BIS had found no way of financing any expansion in their technical work, and World War II, which was a great help for the ARS and made the greatest successes of von Braun's team possible, forced the British group to suspend activities for more than five years. A few members were able to remain in contact, through correspondence and occasional informal meetings, and keep alive a spark that could be rekindled when peace was achieved.

Arthur C. Clarke reports that a private verbal battle was fought to a draw in an Oxford pub, shortly after the war. On one side were Clarke and Val Cleaver, both members of the BIS who had been converted to the spaceflight ideology by the early reading of science fiction.[121] They contended that astronautics would be a good thing, undoubtedly echoing the Society's view.

> Space is a normal and natural [indeed inevitable] development in the history of mankind, comparable to the conquest of the oceans and the air, but greatly exceeding these in scope. Space developments will influence all aspects of human thought in the years to come.[122]

Opposing them were mystical humanists C. S. Lewis and J. R. R. Tolkien, who charged that rocket societies would welcome "the destruction or enslavement of other species in the universe," and that space travel, whether possible or not, was an evil concept.[123] In his novel *Perelandra,* first published in 1944, Lewis said of the chief villain:

> He was a man obsessed with the idea which is at this moment circulating all over our planet in obscure works of "scientifiction," in little Interplanetary Societies and Rocketry Clubs, and between the covers of monstrous magazines, ignored or mocked by the intellectuals, but ready, if ever the power is put into its hands, to open a new chapter of misery for the universe. It is the idea that humanity, having now sufficiently corrupted the planet where it arose, must at all costs contrive to seed itself over a larger area; that the vast astronomical distances which are God's quarantine regulations, must somehow be overcome. This for a start. But beyond this lies the sweet poison of the false infinite—the wild dream that planet after planet, system after system, in the end galaxy after galaxy, can be forced to sustain, everywhere and for ever, the sort of life which is contained in the loins of our own species—a dream begotten by the hatred of death upon the fear of true immortality, fondled in secret by thousands of ignorant men and hundreds who are not ignorant. The destruction or enslavement of other species in the universe, if such there are, is to these minds a welcome corollary.[124]

In 1946 Lewis published another novel *That Hideous Strength,* which describes an epic struggle between the scientific forces of Evil embodied in a research institute, the acronym of which is N.I.C.E., and a few humanistic forces of Good embodied in the character of Ransom, a Christ figure, and in the magic of good, British Merlin.

Patrick J. Callahan has observed: "Lewis would accept Blake's maxim that 'in trying to be more than man, we become less.' "[125] Lewis' evil scientists mutter, "Does it follow that because there was no God in the past that there will be no God also in the future?" They believe that the first space voyages will bring "the beginning of Man Immortal and Man Ubiquitous...Man on the throne of the universe. It is what all the prophecies really meant."[126] Lewis not only argues against the project of spaceflight, but against the concept of space. What lies above us is not space, but Heaven.[127]

The BIS, which had been a purely amateur club with no legal charter before the war, was revived and incorporated the last day of 1945 with a membership of 280.[128] Other British space groups, such as the Manchester Interplanetary Society, founded by Eric Burgess in 1936, were merged into the new BIS.[129] The *Journal* resumed publication and was an important vehicle for the dissemination of spaceflight ideas around the world.

At the end of September 1947 the Society was able to report that membership had grown to 444 and that funds amounting to 734 English pounds had been taken in over the previous year.[130] A study of the membership showed that the average age was 29. Residents of the British Isles accounted for 87% of the members, and 23% lived in the immediate London area. While 13% were students and 19% in miscellaneous nontechnical occupations, 27% were working on aerospace research, engineering, or design. Industrial and research chemists and physicists accounted for 7%. Mechanical, electrical, and civil engineers combined with radio and radar technicians added up to 27 %, and the final 7% were described as other scientific workers. The BIS was no longer pixilated but professional.

In early 1948 P. E. Cleator noted the "melancholy fact that, out of an original 15 Founder Fellows of the Society, I alone survive as a member...."[131] The BIS had become serious business and no longer the home for as great a proportion of fantasts. Like the ARS, it was maturing. The Society came out with a revised design for a moonship, using liquid-fuel technology. Arthur C. Clarke invested his time in propagandizing, and Val Cleaver attempted to create a British missile and space program. A member of both the BIS and ARS, Cleaver has compared the British and American societies:

> The BIS was more of an amateur society, and we relied less on professional help than the ARS. We did not do any experimenting, for we thought—and I am quite sure that we were right—that it was absolutely ridiculous for any society to imagine that it could conduct practical experiments of any significance.... The BIS members organized into groups and did design studies and wrote papers....the BIS formed a platform for speaking and publishing about these concepts....But when it came to actually building hardware and doing the job, it was too expensive for England to undertake.[132]

Membership in the BIS climbed vigorously from 250 in 1946 to 2500 in 1953, then crept upward to about 2900 in 1963, laboring, as one of its presidents said, "under a severe handicap, namely the apathy and antipathy of a large and broad cross-section of the British people towards space exploration...."[133]

The level of the technical articles published in the *Journal* rose, and when its editor analyzed the issues for 1950, he concluded that while 53% of its content was nontechnical, 13% was "borderline," and 35% was thoroughly technical and might be unintelligible to the average reader.[134] The BIS was struggling to become a professional technical space organization in a nation that had little use for one. Unlike the American group, it never allied itself with conventional aeronautical societies and was never captured by the aerospace corporations.

In a half-hearted way, Britain did attempt the development of liquid-fuel engines and IRBMs. In an autobiographical sketch, English rocket engineer C. E. Tharratt tells of his postwar experience:

> I joined the de Havilland Engine Company and there met Val Cleaver who offered me a position with his group of rocket engine designers. So began the happiest design period of my career. What the Rocket Development Department lacked in size, it made up for in enthusiasm, engineering skill, dedication, and the vision and direction of its leader.[135]

In 1963 Tharratt made a strenuous and successful effort to join the "brain drain" to America so he could work on the Apollo program. Cleaver has called the 1945–1955 decade "the 'finest hour' of the BIS," and has described the current British position in space as "undoubtedly...the most unsatisfactory one of all!"[136] The BIS magazine *Spaceflight* reported in 1973 that the cancellation of the Europa II project signalled the end of major rocketry in the United Kingdom and said of Cleaver:

> [His] endeavours to create a major UK rocket industry fell to the divided ambitions of European governments on 27 April 1973. The Rolls-Royce RZ.2 engines for the HSD Blue Streak first stage of Europa I and II, developed under his supervision, achieved a record of 12 successes in 12 flights. But technical excellence was not enough.[137]

In early 1972, Cleaver complained that British investment in space, then $70 million a year, was not only tiny compared dollar-for-dollar with the United States investment, but was only one eighth as great on a GNP basis, and a sorry one twentieth on a per capita basis.[138]

Now the only logical route for European space development is in cooperation with the United States or Russia. The most promising international project currently under construction is the Spacelab module being prepared by the Europeans as a scientific research package for launch by the United States Space Shuttle. Spacelab is costing Western Europe $320 million. The British share is to be only 6.3%, with the major contracts going to Germany which is putting up 52.55% of the money.[139]

Britain's space weakness has been a blessing in disguise for the BIS. Cleaver has complained that although "the BIS is still the only body in the UK wholly dedicated to spaceflight...many of its keenest members are still not professionally involved in astronautics...."[140] A BIS brochure boasts: "Although the Society has close ties with Industry, Government establishments and research organizations of all kinds, it is a fully-independent Body qualified to proffer advice on space research and technology."[141] The BIS has not been absorbed into the aerospace industry as has the AIAA, in great measure because of the weakness of official support for space in Britain. The BIS reports that "Most of the Society's income is derived from membership subscriptions,"[142] and the organizations's 1972 financial report does indeed show that those portions of the £30,000 income that do not come directly from members, derive from the sale of publications and from investments.[143]

The AIAA and BIS have accidentally achieved an excellent division of labor between them. The former is rich, large, and embedded in the industrial-governmental-aerospace complex. The latter is poor. With 3000 members worldwide, the BIS is small, and remains fully as independent as when it was formed. While the AIAA concerns itself with projects showing promise of quick financial gain, the BIS keeps the long view. Although the *Journal of the British Interplanetary Society* is as technical and cautious as the AIAA's *Journal of Spacecraft and Rockets,* its monthly magazine, *Spaceflight,* and its frequent meetings range over the widest spectrum of space topics.

For a few years members of the BIS have been working on a visionary Project Starship to design a craft capable of carrying men "out across immensity to the stars."[144] Recently this group has proposed a Project Daedalus to send unmanned fusion-drive space probes on 40-

year flights to study the system of planets believed to orbit Bernard's Star, 6×10^{13} kilometers away.[145]

Like earlier British designs, this third BIS Spaceship may never be built, but might suggest the features of actual interstellar probes of the twenty-first century. At the 40th anniversary of its founding, on October 13, 1973, the BIS rededicated itself to the Spaceflight Revolution by announcing plans to devote itself to studies of interstellar flight and communication. "The Society says it is time once again to move ahead of contemporary technology to examine new goals in astronautics."[146] Despite its decline in world politics, Germany took us to the moon. Perhaps, in some way, Britain will take us to the stars.

Chapter 6 / THE COMMITTEE FOR THE FUTURE

At the time of its creation the Committee for the Future was the most striking independent organization within the Spaceflight Movement. An amateur group unconnected with the aerospace industry, NASA, or the AIAA, the CFF attempted to inspire support for the colonization of the solar system. In 1972 and 1973 I studied the CFF by attending two of its conventions, through interviews, a questionnaire (Q-CFF) sent to all participants at the May 1973 CFF conference, and its own publications. Over this period the CFF turned completely away from spaceflight and became a pop-futurology group interested in encounter-group-like conventions, unable to contribute to further space development. The course of its history illustrates some of the motives that can inspire interest in spaceflight and some of the barriers to public participation in the Movement.

Birth of the Committee

Barbara Marx Hubbard, one of two daughters of toy tycoon Louis Marx, is the founder, patron, and one might say, *mother* of the Committee for the Future. Born to wealth, she has in recent years been impelled by an inner drive toward fanatic social activism, which was seen by one of my informants as compensation for guilt over her own good fortune. The fact that her sister is the wife of Daniel Ellsberg, the man who made public the famous *Pentagon Papers,* indicates that Mrs. Hubbard's special concerns in some sense run in the family.

Born in New York in 1929, Barbara Marx graduated *cum laude* from Bryn Mawr College with a degree in Political Science. She is proud of the personal journal she has kept since Christmas Day, 1948, and undoubtedly this document will be of great use to some future biographer. The passage she herself has chosen to publish from the very first entry expresses the ever-unfulfilled passion which drives her:

> It is Christmas, but I feel none of the mystery, the peace or the warmth. All the beautiful feelings which come to one on Christ's birthday shun me. Instead I am tortured by doubts, fears and unhappiness. There is a constant pull in the middle of my stomach. I am torturing myself to death. The cause is evident: In my own eyes I have achieved nothing, yet those same eyes have visions of glory untold. There is a key to my desires which I hold but cannot use. I must either lower my ideals or achieve them. I am like a magnet feeling the attracting force of another magnet, yet held apart. It's tearing my insides out.[1]

Barbara spent her junior year of college in Paris, where she met her future husband, Earl Hubbard, who told her "he was an artist in search of a new image of man commensurate with our new power to shape the future."[2] Interestingly, she says she "fell in love with that aspiration" and married Earl a year later, but she does not say she fell in love with the man himself. The Hubbards settled in a palatial house in Lakeville, Connecticut, and devoted themselves to producing five children and many paintings, living, as Barbara put it, on the "welfare...of a generous father."[3]

She says that throughout this period she frequently enjoyed what her later friend, Abraham Maslow, has called "peak experiences."[4] Barbara was gripped by a profoundly moving peak experience, almost a religious conversion experience, in 1967, partly in response to Earl's invention of a space-oriented painting style. He gave her a drawing of her face, gazing upward, rendered in this bold style. After naming it *Mankind in the Universe,* she took a long walk, brimming over with emotion. It was as if she had been transfigured. There were no longer any limits to her hopes, and a rapturous sense of Destiny drew her onward.[5]

She immediately set herself to the task of translating Earl's emerging philosophy into concrete form. Working with a tape recorder, she transcribed their breakfast conversations and edited them into a book and several articles, some of which have been published under his name. She compared their efforts to those of the authors of the Gospels who were privileged to announce Christ's birth, and she was struck by the idea that she also had a birth to proclaim—that of Man into the Universe. It was a rebirth for herself as well, and in this stunning vision of cosmic glory she found "that meaning which I needed to live."[6]

The mataphor of birth appears frequently in the rhetoric of both the Hubbards, and Barbara continually seeks to participate in the births of new movements and processes. She does not fix her efforts steadfastly on any one project, but rather chases the fleeting sensations of peak experiences, and therefore must forever be launching grand beginnings, instead of merely plodding along in on-going projects.

Earl Hubbard's "space philosophy" is essentially assertive, moralistic, optimistic, and is not Philosophy in the modern, academic, critical sense. He sees his function as presenting a positive vision of life and of Man's future, an image that can inspire and guide masses of people. Although he makes a variety of claims, two basic principles seem to tie everything together in his thinking:

1. Integration of all individuals into a unified body of Mankind, what Teilhard de Chardin called "convergence"[7]
2. Transcendence of the human condition through evolution into a "universal species"

In political terms, the first calls for world government, world monitoring, and the development of an unchallenged consensus. The second calls for an aggressive space program. In psychological terms, the first suggests the possibility of a thorough experience of fusion with other persons and incorporation into the single person of a monolithic "Mankind." The second urges ascendancy, whether in the form of increased individual self-esteem, in liberation from the oppression of gravity and the gravity of oppression, or in an evolutionary progress into higher and better modes of being. The political ideology may be seen as a projection of the psychological longings onto the external social and material worlds.

Earl Hubbard is happy to see his purpose as essentially religious. At one point he calls for the formation of a new sacred organization, the Cultural Cathedral, "to give us a cosmic perspective and to serve as a central point of concern for Mankind, its future in the universe, its relationship with the creative force."[8] This new religion will not be at odds with science; rather, it will include science, which Hubbard sees as little more than an extension of the search for sacred meaning: "Out of prayer was born science. Pure science is to seek the intention of creation."[9] He calls modern science an "ecstasy of revelation,"[10] and asserts that it is our duty, through science and spaceflight, "to search for the intention of creation"—that is, to search for God.[11]

He believes biological and cultural evolution are expressions of the divine will which now demands our conscious assent and active participation. "Mankind is evolving into the universe. This is where the new challenge lies. Without that objective the meaning will drain out of every organization in existence, including the human body."[12] Hubbard speaks of mankind as "a theotropic form of energy,"[13] a God-seeking spirituality. The individual man is also bound to evolution, is theotropic, and has the duty to ascend into the heavens: "The individual is Mankind."[14] Hubbard assigns to the individual and to the species the mission of arising and evolving on behalf of the entire planet, and, like Bergson,[15] even makes the sun a part of our constituency:

> Man's task on earth has been the uniting of all known energies to overcome all known obstacles to life. And now man has created a means of reaching beyond the community of earth. We represent the possibility of carrying a seed from this system elsewhere. We are a means of union between this system and a larger system. Man appears to be the agent of the transformation of the material forces of earth into a transcendent force. Man is the agent of earth's transcendence.[16]
>
> ...all action on earth has been part of one Creative Intention to convert the sun's energy into a form capable of moving out from the sun into the universe. Everything about the earth appears animated by this intention. Everything that has happened on this earth appears to have been essential to make leaving this earth possible.[17]

Two anthropologists have contributed further mystical ideas along these lines to CFF ideology. Dr. Roger W. Wescott of Drew University has told a CFF convention audience that Man is "preadapted" to space. Preadaptation is a principle rather like the sociological concept of anticipatory socialization, but without any rational explanation. In biological evolution, Wescott believes, a species sometimes acquires characteristics that prepare it to advance to a new ecological niche in the future. Early amphibians had the ability to breathe air; men have the ability to work in zero gravity. Clearly this concept is part of a teleological theory of evolution; spaceflight is part of God's plan and He has prepared us for it.[18]

Egyptologist Dr. Walter Fairservis of The American Museum of Natural History, known for its deviant and Humanist brand of Anthropology, has written two plays for performance at CFF meetings. *Marut: The First Thousand Years,* is a somewhat liturgical translation of *Oedipus Rex* from ancient Thebes to the future colony Marut on a distant planet. *The Pyramid* is an allegory of Man's struggle to evolve told in terms of a small group of people who are trapped in an ancient Egyptian pyramid and must find their way up and out to salvation. One of the characters becomes lost in deep passageways; as he gropes for the way upward, he experiences not only his own escape and uplifting, but also the entire progress of human history "from the primordial world where the living prey on the living and struggle in darkness moved only by appetite...towards Heaven."[19] The imagery of this scene combines two occult traditions. One is the absurd notion that the ancient pyramids were built by a master race which inscribed all past and future history in the design of the twists and turns of the passages.[20] The other is the discredited theory propounded by Ernst von Haekel that "ontogeny recapitulates philogeny," that in the stages of individual development can be seen the stages of species development. Rosicrucian tradition strongly holds that there is a unity of the evolution of mankind and the salvation of the individual man; this idea is implicit not only in the ideology of the CFF, but also in the theory of evolution put forth by rocket pioneer Oberth.[21]

From 1967 to 1969 Barbara Hubbard sought to realize her ideals of human unity and transcendence and to join with other like-minded persons through the medium of a publication called *The Center Letter.* She began by writing letters to about 1000 prominent people concerned with "the need for a better future," "asking them what they thought was the next step forward for the future good." Many responded with letters or literature, and she digested and reprinted their comments in her newsletter. She says she was "overjoyed by the immense force of good-will in the world....At this time she became a prominent member of the World Future Society, a Washington-based pop-futurology organization with about 10,000 members and a monthly magazine. Through these activities she developed personal contacts with a large number of people, which she could draw on later to flesh out the CFF; for example, at least 13 of the 80 CFF participants who answered my questionnaire had first learned of Barbara's group through the World Future Society.

The sixth issue of the WFS magazine, dated December 1967, carried on its cover a reproduction of Earl's painting *Mankind Alone* and contained Barbara's philosophical-ethical article "Mankind in the Universe."[22] She was a prominent participant in a WFS leadership meeting held on December 12, 1967,[23] and by the next issue of the magazine had become its "Images of Man" editor.[24] In 1968 the WFS printed three brief articles by her, and she and her husband contributed frequently thereafter. In an interview with Ed Cornish, founder of the WFS, I learned he and the Hubbards had become very close and shared a number of personal friends.

In June 1970 the Hubbards invited a number of associates to their home to discuss the formation of a future-oriented organization of their own. Included in the group was Lady Malcolm Douglas-Hamilton, said to have organized the World War II "Bundles for Britain" campaign and founder the the Center of American Living, of which Barbara was an executive and to which Earl had lectured.[25] At least four aerospace professionals participated in the gathering—General Joseph S. Bleymaier, who had helped direct the Titan III-C booster program; Harold W. Ritchey, chairman of the Thiokol Chemical Corporation; James Sparks of the Boeing Company and author of the book *Winged Rocketry*;[26] and Lt. Colonel John J. Whiteside, former Air Force information officer and adman who came to dominate the Comittee after 1971. A religious dimension was added by the presence of Sister Mary Fidelia of the Felician Sisters, a future-oriented Catholic activist who often quotes Wernher von Braun, who speaks of the "marriage between 'Sister Faith' and 'Brother Science,' " and who became a close friend of the Hubbards and supporter of their projects.[27]

On June 21st, the group of 15 proclaimed the birth of the Committee for the Future through a document called the Lakeville Charter, which states:

> Earth-bound history has ended. Universal history has begun. Mankind has been born into an environment of immeasurable possibilities.
>
> We, The Committee for the Future, believe that the long range goal for mankind should be to seek and settle new worlds. To survive and to realize the common aspiration of all peoples for a future of unlimited opportunity, this generation must begin now to find the means of converting the planets into life support systems for the race of man....

> We believe that the time to state the new goal is now. Awareness of the new option can transform this troubled world from a place of despair into a sphere of hope, aspiration and joy.[28]

Drawing on Mrs. Hubbard's social network, the CFF gained in strength, and at the September 1970 Third Lakeville Conference on New Worlds, 32 people subscribed to a second proclamation in which they affirmed their "commitment to strive towards a liberation from the bonds of the earth to exploration of new worlds" and announced that their purpose was "to reach out, change dreams to reality, and be born as new children of the universe."[29]

Project Harvest Moon

The most interesting space effort of the CFF was the abortive Harvest Moon Project. This plan called for an international, perhaps civilian *Apollo* flight to perform experiments and demonstrations that would lay the groundwork for later colonization and economic exploitation of the earth's satellite.

Committee member George Van Valkenberg, a TV film producer, suggested the idea in October 1970. He pointed out that NASA had cancelled two moon flight missions and that, therefore, hardware was available for a privately financed venture. In January 1971 the New Worlds Company was incorporated, "financed with a loan from the Louis Marx Company," to develop Harvest Moon. The officers of this company were Mrs. Hubbard, her friend Margaret Morrow, and John J. Whiteside.[30] Throughout the year the CFF explored possible sources of financial support and developed a general plan for the mission.

The cost was estimated at $150,000,000, assuming that NASA would donate the *Saturn V* launch vehicle, *Lunar Lander,* and *Apollo* modules. The crew was to include a Russian cosmonaut to give the undertaking the proper international flavor. A landing would be made at Hadley Rille, the site explored by the *Apollo 15* mission, and which is without doubt the location with the best tourist scenery of those places on the moon yet visited. The CFF flight could reuse some equipment left over from *Apollo 15.*[31]

The main focus of the astronauts' work would be deploying an experimental package called FIELD—First Integrated Experiment for Lunar Development,[32] which would consist of four parts:

1. A prototype lunar garden under a plastic dome
2. "Roger," a robot lunar rover
3. A laser communications relay system
4. A small telescope, "First Lunar Observatory"

The information available about these pieces of apparatus is very sketchy, and I have no reason to believe that detailed designs or technical studies were ever made. We know that Roger the robot rover would have been feasible, because the Russians have actually operated two similar Lunikhod vehicles, but the CFF drawing of Roger is very crude. The laser communications system has three drawbacks: (1) It could hardly compete with the already-existing array of communication satellites; (2) it would require laser technology considerably advanced over the current state-of-the-art; and (3) there seems to be no provision in the device for a power supply! The lunar telescope would be vastly inferior to the large orbiting telescope being developed by NASA but would have been an amateur brother of the actual ultraviolet telescope used on the moon by the *Apollo 16* crew.[33]

The lunar garden experiment was a wonder of nontechnical, romanticist, rhapsodic thinking. In this plan the astronauts would inflate a 20-foot mylar dome over lunar soil, insert plants and small animals, switch on compact life-support equipment, and return to Earth. The progress of the budding flora and scrambling fauna would be observed from Earth over television. Supposedly, "atmosphere would be provided by the residual consumables in the LEM." What consumables? The descent stage fuels were toxic, so they could not fulfill this function, and the ascent stage would have to lift off the moon quickly. Clearly, no residual consumables were available. Furthermore, what would keep the dome from hopping up, lifted by air pressure, and spilling its atmosphere into the lunar vacuum was never explained. Nor was there any provision to shield the dome's lifeforms from the deadly temperatures of the boiling day and freezing night. That Roger would prospect for water was vaguely suggested.

Harvest Moon's "long-range objective [was] to stimulate governments to cooperate in establishing the first lunar community as a staging ground for the human role in the universe."[34] In its promotional literature, the CFF stressed the project's value both as a means for fostering world unity and as a step toward man's colonization of the universe. As a scientific or technical enterprise, Harvest Moon would be meaningless; it was intended as a symbolic act with social and spiritual meanings.

Initial contacts with NASA did not provide encouragement for the project, but throughout 1971 the CFF carried on a campaign to locate interested supporters. A promotional film was made and shown to a variety of small audiences; Northwestern University engineering student and space enthusiast Carl Konkel carried it with him to a space conference in Brussels. Editor Norman Cousins was briefly interested in Harvest Moon and held a small seminar about it in November of that year.

On March 9, 1972, leading members of the CFF testified in favor of an expanded space program before the Committee on Science and Astronautics of the U.S. House of Representatives. They described Harvest Moon as a "venture for humanity by humanity initiated by individuals of all races, nations, and religions who want to secure for themselves and mankind an unlimited future," and they asserted that their intention was "to open the frontier on new worlds for all mankind, beginning with the moon."

Perhaps the highpoint of enthusiasm for Harvest Moon was during a European promotional junket conducted by Whiteside and Mrs. Hubbard in April 1972.

Texas Representative Olin Teague, a long time space booster who had been in communication with the CFF for over a year, introduced a House resolution on May 11, 1972, calling for NASA to donate the necessary space vehicles to the CFF for Harvest Moon. The Committee launched a halfhearted effort to organize public support for the resolution. By early 1972 Earl Hubbard's involvement in the CFF had ended altogether, and the group was actively preparing the first of John Whiteside's "Syncon" meetings. CFF newsletters said Harvest Moon and Syncon were its two major projects, of equal importance, but clearly the latter had become the real focus for action.

At the May 1972 Syncon convention in Carbondale, Illinois, George Robinson, the Assistant General Counsel of Washington's Smithsonian Institution, who was serving without pay as the Committee's lawyer, told me that much of his current work involved setting up the necessary legal structure to permit the large-scale raising of funds for Harvest Moon. He said, "A rather interesting and innovative legal complexity is going to have to be developed in accordance with present plans to solicit funds and not incur liabilities. What do you do if you don't come up with a project and you have to give your money back?" By this time Robinson himself was one of the few CFF leaders deeply concerned with Harvest Moon. Mrs. Hubbard had apparently been lured away from space by Whiteside's Syncon. Her only public activities concerning space were conversations and proclamations about the possibility of financing a television series about a lunar colony to be produced by Gene Roddenberry, the creator of the Star Trek program.

Carl Konkel was appointed leader of the Carbondale Syncon Space Task Force, and as part of the wide-ranging discussion of the future of the space program, Harvest Moon was considered as a plausible step toward a moon colony. It was suggested that an experiment be added to the plan to test Krafft Ehricke's hope that underground explosions might liberate useful quantities of oxygen from the lunar rocks. The group's "Timeline for Lunar Development" did not, however, actually include Harvest Moon. Konkel and his team of engineering students made a formal proposal for what they called Project Skylink, an international second Skylab earth-orbiting laboratory which would be visited by both Russian and American crews. They brought with them a nice model of Skylab B being docked by a white *Apollo* and a light blue *Soyuz.*

On July 18, 1972, Olin Teague called a meeting of NASA representatives, Mrs. Hubbard, and Whiteside to discuss Harvest Moon. Apollo Director Rocco Petrone told the saddened CFF delegation that the stored lunar lander had been "cannibalized for parts" and was in no condition to be sent to the moon. After considering the situation, the CFF proposed that Konkel's Skylink be substituted for Harvest Moon under the evocative name, Mankind I.[35]

There were spasmodic announcements of the Mankind I project in CFF newsletters, but apparently no major planning or publicizing campaign got under way. According to the first newsletter of 1974, "CFF hopes in promoting transnational space project ran out of steam in the face of bureaucratic delay." Interestingly enough, the CFF has shown no interest at all in the real transnational Apollo-Soyuz Test Program, announced by the United States and Russia at the end of May, 1972.[36]

Abandonment of Space

The shift in emphasis away from space toward a focus on world planning and pure sociation had already begun in May of 1971 when the CFF held a conference at Southern Illinois University at Carbondale. The Committee had developed a relationship with various members of Buckminster Fuller's pop-futurology World Game organization, which had its headquarters at the university. Fuller himself was present at the meeting, and an element of his philosophy crept into CFF thinking, particularly the convergent, mystical notion of "synergy." The featured speakers were Earl and Barbara Hubbard and John Whiteside. According to an informant, several of the announced speakers failed to show up, the meeting was very disorganized, and a power vacuum was created into which a variety of individuals attempted to interject themselves.

At this meeting Thomas Turner, Director of the Fuller Projects Office of S.I.U., made the formal proposal that Barbara Hubbard seek the 1972 nomination of the Democratic Party for President of the United States; with one dissenting vote, the conference adopted this as a resolution.[37] Turner has identified himself with Earl's space philosophy, saying:

> The new worlds movement is NASA finding its soul....The testimony of the Committee for the Future reiterates that God's work is truly our own rationale for existing. It has tenaciously pursued its insights knowing that as long as it works with evolution it cannot do harm. NASA, unknowingly, has been working with evolution but now is stalemated when the embryo obviously cannot differentiate without something more unique than mechanical efficiency—it needs a "soul."

In early September 1971 the New Worlds Labor Day Conference was held at the Hubbard home in Lakeville, Connecticut, with about 50 people in attendance, including John Whiteside. This is the last major meeting Earl Hubbard attended, and since then Barbara's escort has been Whiteside. This meeting, like the one a year earlier, laid initial plans for a Carbondale conference to be held the following May. It was to be a conventional gathering, with a variety of panel discussions and other activities. Perhaps chastened by the 1971 fiasco and inspired by innovations attempted by other futurology groups, in December Whiteside devised a new structure for the meeting, called Syncon from the phrase *synergistic convergence,* a union of Fuller's and Teilhard's concepts of creative human union. At the precise time that Whiteside's ideas were adopted, Earl Hubbard dropped from the picture.

I have not seen any CFF documents that include the name of Earl

Hubbard among those described as full members of the Committee. Throughout 1971 the masthead of *New Worlds* magazine proclaimed: "Earl Hubbard, artist and space philosopher, is special advisor to the committee." The last issue bearing that legend, also the last to contain an article by Earl, is that for December 1971. The January 1972 issue announces the birth of Whiteside's Syncon and is the first to drop Earl's name from mention.

In May 1972 at Carbondale I asked many persons where Earl was and why he was absent. Several said they expected him to appear and were surprised when he did not, although none of the publicity for the event mentioned him. According to one informant, Mr. and Mrs. Hubbard had "a philosophical falling out in terms of how his philosophy was being used." Earl was said to be opposed to a highly structured approach and did not approve of the public action programs the CFF was getting into. There is no evidence he ever participated actively in either the Harvest Moon or Syncon efforts. The Hubbard's housekeeper, perhaps attempting to play down the significance of the rupture, claimed that Earl had remained in Lakeville to work on a new book: "You know writers. When they get working they want to get away. They don't want their thoughts interrupted." This is implausible, because Earl's method of writing was to chat with Barbara about his ideas over breakfast and have her produce the written manuscript from notes and tape recordings. In any case, no such new book has appeared.

An important symbol of Earl's influence in the CFF and in Barbara's life is the portrait of her he drew in 1971. Until the end of 1972 this image of Barbara's aspirations as interpreted by Earl was the logo of the CFF. The picture was reproduced in every possible form for the Carbondale Syncon—buttons, stickers, posters, literature, name tags, wall decorations, and even on a huge floating balloon. To explain the style of this painting, *Mankind in the Universe,* Earl has said: "The background of man is no longer earth. It is space, and the color of space is black."[38]

I noticed that Barbara wore a ring with this picture in black enamel and silver and asked her about it. She said that a Los Angeles jeweler had become interested in the goals of the CFF, saw the logo, "suggested doing it, and so John [Whiteside] got it for me for my birthday." Whiteside, almost always at Barbara's side, interjected, "If people like them well enough, we'll just put them out for sale." Barbara commented, "It's the face that launched a rocket." "Thousand rockets," Whiteside corrected.

When the CFF later moved to Philadelphia and had new stationery printed, Earl's face logo was replaced by a black and white bull's-eye design. Over the same span of time, the mottos on the head of the stationery changed, reflecting a transformation in the interests of the CFF:

1971: Proposing the development of the frontier on new worlds in space for all makind

1972: Proposing new worlds on earth—new worlds in space

1973: A non-profit organization dedicated to bringing the options for a positive future into the public arena for decision and action.

1974: Building a positive future [The motto printed by the Committee's postage meter.]

In June 1972, Barbara and some associates moved the CFF from Lakeville, Connecticut, to Philadelphia. Some of the local staff, including New Worlds Company officer Margaret Morrow, resigned. CFF announcements predicted that a permanent on-going Syncon would be set up there and that it would become the focus for traveling Syncons that would venture forth to all corners of the earth. Of course this move represents, in concrete fashion, Barbara's estrangement from Earl and from his space philosophy, and her complete adoption of Whiteside's goals. After an indecisive period, in the fall of 1973, the CFF set up a New Worlds Training and Education Center in a palatial 18-room house in Washington, D.C., which in effect became the organization's new headquarters and a staging base for Syncons.

Perhaps the last pro-space undertaking of the CFF was the Delta Project, an attempt to find optimistic alternatives to the pessimistic findings of the Club of Rome's *Limits to Growth* study.[39] At the May 1972 Syncon a young computer worker, Ray Frenchman, gave a simple exposition of this study, explaining that students of MIT's Jay Forrester had applied his world modeling techniques to an analysis of future resources and the quality of life, coming up with dire predictions. Frenchman said that this report should not be taken as the final answer and urged that the questions be investigated further. At this Syncon, North American Rockwell rocket expert and former Peenemünde team member Krafft Ehricke made an impassioned address, calling for the commercial exploitation of space and of the moon, concluding:

> It is painfully elementary to conclude that unlimited growth is not possible in a limited, closed world. It is equally elementary that an open world—while not permitting unlimited population growth due to the present uniqueness of this planet—does, however, permit the establishment of a basis for the material growth needed to provide an adequate living standard for the inhabitants of this planet without destroying its environment. This growth...makes possible the continuation and deepening of civilization, no longer interrupted periodically by retrogression into Dark Ages.[40]

Ehricke's biographer says he has always been fascinated by "the great trends affecting the rise and fall of civilizations."[41] When I interviewed him, I found that he was gripped by Oswald Spengler's theory of the natural growth and decline of cultures and believed that expansion into space could change the dreary laws of history. The idea that exploitation of space could remove any economic limits to growth was presented in a detailed analysis as early as 1931 by the Russian, N. A. Rynin, although most space scientists have been skeptical of the feasibility of this option.[42] A few hours after Ehricke spoke, John Whiteside announced that one of the CFF's four current projects would be "The Opportunities for Mankind," the answer to "The Limits to Growth." This effort was renamed the Delta Project and was discussed by Mrs. Hubbard later in Bucharest with Club of Rome founder Aureilo Peccei; she "presented [it as] our challenge."

The CFF hired Jon Michael Smith, a McDonnell-Douglas employee, later described as president of The Software Research Corporation and a member of the CFF's Research and Development Advisory Group, and gave Delta over to him. He was scheduled to give a presentation on Delta at the October 1972 Los Angeles Syncon. For a time the CFF offered members a $1.50 interim report on Delta, a slick uninformative bit of advertising confection. In it there is no mention of the hope that space development might play any part in the world's future, nor any suggestion that economic expansion can continue without limit. No mention of Delta was made at the May 1973 Syncon,[43] and by January 1974 it had been "incorporated in the ongoing activity" of the Committee, that is, aborted.[44]

In the Washington meeting, Extraterrestrial task force leader, George Robinson, complained about the CFF deemphasis of space and new concern with transforming the earth into a "stable platform" for Man: "I have a feeling they're getting a little hung up on stabilizing earth as a platform first." His friend, fellow panelist, and aerospace author, Michael Michaud, agreed: "It's wrong to wait for this platform to stabilize." Robinson, it will be remembered, was the CFF's attorney and organizer of part of Harvest Moon; it was correct for him to speak of the CFF leadership as "they," because apparently Whiteside and Mrs. Hubbard did not consult all their lieutenants about the major shift in direction of the organization.

Synergistic Convergence

Whiteside's Syncon process turned the focus of the CFF from outer space to inner space. It is a think-tank procedure that seeks to achieve a creative "synergistic convergence" among numbers of ordinary people who come together to plan the future of the world. The physical layout of several specialized discussion groups and the channels of communication between them are progressively manipulated over the space of three days to produce a growing consensus, a unified community of interests, outlooks, and accepted opinion. It brings people together ideologically by bringing them together physically.

The heart of the physical environment is the Syncon "Wheel," a circular structure of room partitions divided like a pie into six sections, each of which holds a task-force discussion group. Around the Wheel are other spaces set aside for still more groups. The chart accompanying this chapter shows the layout of the Washington Syncon, with the various areas labeled.

Synconners are very definitely not experts in relevant fields, nor are they policy makers. Anyone willing to pay the registration fee can participate fully. Each person signs up for one of the basic discussion groups and begins his Syncon experience in the part of the structure set aside for his task force. Each group has an informal leader, briefed by the Committee at an earlier meeting, and a list of suggested questions to take up. The agenda and the power structure may be negotiated in any manner agreeable to the participants. Whiteside is opposed to the incorporation of standard futurological techniques into the simple Syncon process; for example, he has discouraged the use of Delphi methods.

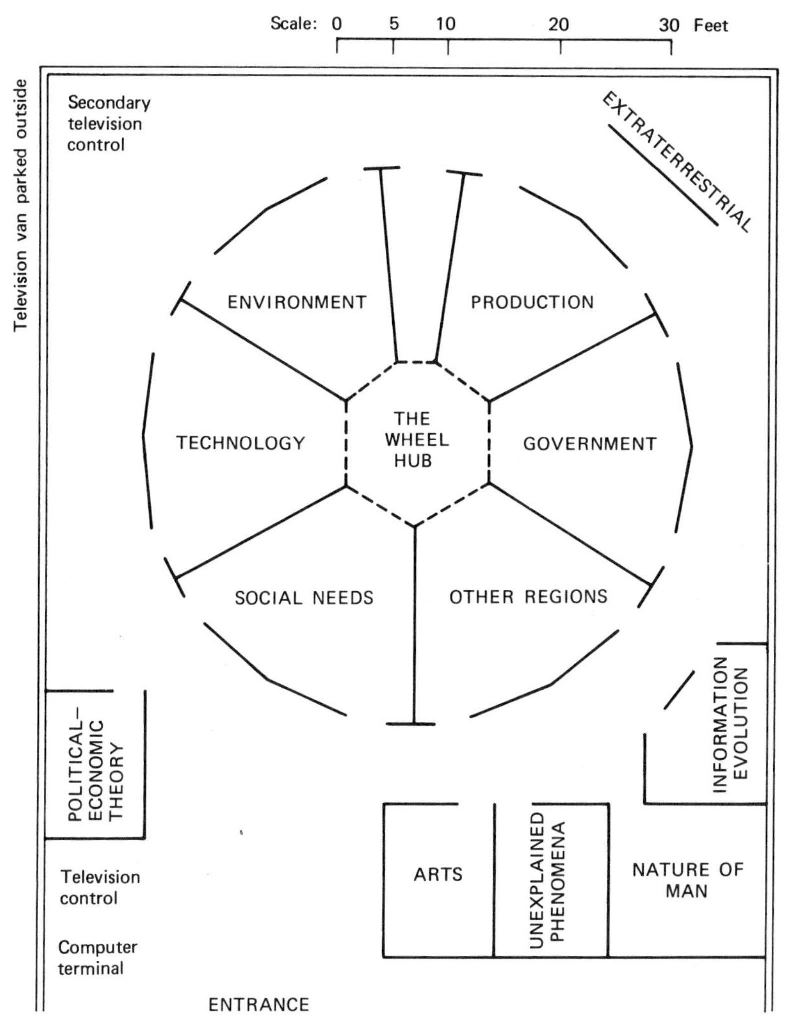

Illustration 6.1 Chart of the Syncon Wheel and satellite areas in the ballroom of The Sheraton Park Hotel, Washington D. C., May 7-10, 1973.

The processes in the groups follow one or the other of two patterns. Sometimes, as in the Washington Syncon's "Social Needs" section, a dynamic leader will manage the various forces in the group and bring about a quick consensus followed by effective planning. In this case the leader took charge immediately with a questionnaire asking participants to decide on an agenda and rank various goals in terms of importance. At the opposite extreme, groups will wallow in the time-consuming anomie of a newly formed T-group, until a painful and emotional sense of communion has been forged. After such an experience, the "Government" group posted the following notice to nonmembers: "Friends: We welcome *observation*—but—because of the emotional dialogue which occurred this A.M. we cannot honestly accept new participants at this time."

Despite their amateur composition, each task-force group did in fact treat the others as panels of experts. Whatever reservations the individual participants may have had, as a cluster of groups bound by norms of respect, they did play the game. Often a group would come to an impasse in its discussions and pass the buck, in the form of a written question, to another group, just as they had been instructed to do. Whenever possible I observed this process, collecting the messages and early drafts of them, and attempted to reconstruct the sequence of communications. For example, the "Applied Technology" group sent a message to the "Extraterrestrial" group attempting to define their relationship:

> As we see it, division of responsibility between you and us as regards XT [Extraterrestrial] colonization, exploration, etc., is: We examine technological constraints associated with these activities; you evaluate the impact of the events.

The spokesman for the Applied Technology group was an Air Force rocket engineer, whom I had interviewed at length a year earlier and naturally took it for granted that the Extraterrestrial task force would be solidly in favor of a vigorous thrust into space. There was a division in Extraterrestrial, however, and it replied: "We are still debating whether humanity needs to explore and colonize on other worlds."

The leader of the Extraterrestrial group said repeatedly that it was man's nature to explore and that in the preordained process of evolution, humanity would naturally have to explore and colonize other worlds. This was disputed by two of the seven members, and the disagreement could not be resolved within the group because there was no agreed-upon process for settling disputes. One member suggested a vote, but this idea was not accepted because an overt division would threaten the development of complete unity. The matter was resolved by passing the question to the "Nature of Man" group in the following message: "Is it the nature of man to expand from the earth into the universe?"

The Nature of Man task force wrote the question on its blackboard and discussed the issue. At one point they considered the following reply: "It is inevitable that Man expands from earth into the universe." But members of the group suggested amendments, and consensually they decided on the following answer, which was sent to the space group: "It is the Nature of Man to experience both outer and inner space. P.S. There is a basic exploratory drive."

At the same time, Applied Technology was communicating about a similar question with The Nature of Man. One Technology message inquired:

> What part (percentage) of our needs are a result of "programmed desires" and "programmed needs"? Background: We are interested in understanding the degree to which the generation of material goods (energy absorbing and pollution generating) is really required to meet human needs.

The Nature of Man debated this and sent Technology this definitive reply: "Approximately 80% of our needs are the result of cultivated desires and cultivated needs." Technology sent a copy of this exchange to Extraterrestrial, apparently considering it relevant to the earlier discussion. Another connected query Technology sent The Nature of Man was:

> It is the nature of man to be a parent? Background: This question arises in relation to the energy crisis. The energy problem is *directly* related to population growth. We are interested in understanding the fundamental drive to generate people.

The ambiguous reply from The Nature of Man was: "It is the Nature of Man to continue the species by (A) producing children, (B) not producing children as circumstances/necessities require."

On the second day of Syncon the various groups are supposed to come up with general recommendations about the future of the world, or reports about our current condition, which are announced to the entire body by spokesmen over closed-circuit television. Members in each section nervously prepared for their presentations while they watched the others on the ubiquitous TV monitors. Mobile TV cameras entered each section of the Wheel.

After the reports, the first "Walls Down" ceremony was held. The partitions separating some sections of the Wheel were removed, and pairs of task forces united into larger groups with wider discussion mandates.[45] In Table 6.1 I show the results of two censuses I made of participants in the different task forces, before and after the first Walls Down at the Washington Syncon.

We notice that some attrition occurred over the space of one day. At the later All Walls Down ceremony, in which everyone came together, again 103 individuals were present. The convention membership list included 249 named participants, but I know that several stayed for only a short time, and undoubtedly, others did not actually attend at all. The convention hotel provided 220 plates for the buffet lunches and 210 places at the dinner tables, and nearly this number did partake of the meals. The difference cannot be accounted for by the free-floating CFF leadership or the TV crew, and I presume some participants went sight-seeing in Washington, were busy with their regular jobs, or

TABLE 6.1 CENSUS OF PARTICIPANTS IN WASHINGTON SYNCON TASK FORCES

	Tuesday 10:15 A.M.		*Wednesday 1:30 P.M.*	
Task Force	*People*	*Percent*	*People*	*Percent*
Technology	7	6.8%	→ 14	18.7%
Environment	12	11.7%	↗	
Social needs	12	11.7%	→ 10	13.3%
Other regions	5	4.9%	↗	
Government	10	9.7%	→ 15	20.0%
Production	7	6.8%	↗	
Unexplained phenomena	13	12.6%	→ 29	38.7%
Nature of man	15	14.6%	↗	
Arts, etc.	6	5.8%	→ 7	9.3%
Information	6	5.8%	↗	
Political-Economic	3	2.9%	—	
Extraterrestrial	7	6.8%	—	
Total	103	→	75	

stayed in their hotel rooms. In any case, the meals were more popular than the seminars.

On the third day at the All Walls Down ceremony, the partitions dividing the Wheel were removed, and participants came into the center from the outlying satellite rooms. For a number of Synconners this event was an important emotional experience. When I asked Mrs. Hubbard which moment of the Washington Syncon most stood out in her mind, she replied: "The spontaneous celebration of the Walls Down." A choir from a local Catholic school sang, and "spontaneously" people clasped hands and circled around the Wheel, producing, in the words of various participants: "a strong spirit of oneness," a "very moving" "final convergence," a "great positivity," a "community song/dance procession," and a "snake dance." While one person said "The very end of Syncon when everyone joined hands and chanted 'Om' was a very powerful and spiritual moment," another complained about "people dancing around when there were many problems to be solved."

Barbara Hubbard has said that "the Wheel is a model of the whole system of mankind as one integrated body," and from this perspective the All Walls Down moment represents the final fusion of individuals into a single person. Indeed, Barbara has compared Syncon to a newborn baby, "since my model is the model of birth," and identifying the various task forces as the organs of the body, said that to work together as a whole, "some kind of walls must come down between those organs."

The Occult Putsch

Between the 1972 and 1973 Syncons the Space, or Extraterrestrial, task force lost importance, while the occult Nonverified, or Unexplained Phenomena, group gained.

For the Carbondale 1972 Syncon, the conceptual layout of the Wheel was modeled after the earth-moon system. The six main task forces, including Space, filled the central, terran Wheel, surrounded by an atmosphere of lesser groups. The conference's promotional literature showed the Nonverified Phenomena seminar as a satellite existing apart from the Wheel, with no connection to any other seminar. The discussion topics for this deviant group were listed as: "Paraphysics,

Archeological-Cosmic Mysteries, Expanding States of Consciousness, Parapsychology," and "the Para-Normal field."

A great hue and cry went up from these "nonverified people." First, they complained that their group had been exiled to a distant part of the S.I.U. student center building, 100 walking yards from the center of the Wheel, along twisting corridors. They moved themselves defiantly into a corner of the common room immediately adjacent to the Wheel. Second, they objected to the "nonverified" label. A member crossed out the *non* on the sign identifying the group. At the October 1972 Los Angeles Syncon, it was still called "Nonverified," but by May 1973 the group had been upgraded to "Unexplained Phenomena."

According to one long-time associate of the CFF, the Nonverified group was not originally desired by the leadership, but had been forced on them by the occult-leaning, leftover members of Buckminster Fuller's World Game organization, their hosts at Southern Illinois University. Furthermore, when the 1971 "Man in the Universe Week" was held by the Hubbards at S.I.U., many of the scheduled hard-science panel members had failed to show up, leaving the field to anyone who wanted to take over. A group of people interested in the occult organized around NASA employee Jim Beal, who had a strong interest in such matters, and held their own counter seminar. Beal became the head of the Nonverified/Unexplained groups at later Syncons.

At a mass meeting of Carbondale Synconners, Nonverified member Jean McArthur, who calls herself "an American sensitive," made a tearful and impassioned plea for her orientation:

> To conquer outer space, you have to conquer inner space, first. And that's within all of you. You can have all the think-tanks you want; you can have a government that tells you this, and an educational system that tells you that, but until you can get out of a two percent and a one percent of the mind and get up to the higher consciousness within you, all of the God quality within you, we're in trouble.

That Friday rumor spread throughout the Syncon that Mrs. Hubbard and Whiteside were enraged at an article in that day's issue of the university newspaper, because it headlined "Psychic phenomena called chief aspect of Syncon convention study." The article was entirely based on an interview with Jean McArthur, who called her group "the most important aspect of the entire Syncon convention."[46] Indeed, the nonverified folk constituted the largest single group—15.5% of the total participants.

Astronaut Edgar Mitchell, the lunar module pilot on *Apollo 14,* addressed one of the convention's dinner meetings, saying that the view of Earth from space gave one a new perspective on our world and on man's place in the universe. The next day he bypassed the Space task force completely and became an active member of the Nonverified group. A few months after the Syncon, Mitchell left the astronaut corps to devote his full time to the study of ESP and other psychic phenomena. This was hardly a new interest for him; he had actually conducted telepathy experiments on the way to the moon.[47] He has taught a course on parapsychology at Kent State University[48] and has donated time and money to the study of Israeli alleged psychic Uri Geller.[49] I do not wish to suggest that space makes mystics of men, nor that an interest in outer space inevitably leads to an interest in inner space, but other astronauts have become similarly interested in religious activities and the occult. James B. Irwin of *Apollo 15* has recently been stumping the country as an evangelist in Youth for Christ,[50] and *Apollo 9* crew member Russell Schweickart has made public appearances on behalf of the Maharishi Mahesh Yogi's Transcendental Meditation.[51]

The occult got a great boost on May 8, 1973 when, according to a CFF bulletin, California plywood tycoon and "philosopher" Irving Laucks donated $100,000 to "endow the Unexplained Phenomena Section of the Syncon Wheel." At the May 7–10, 1973 Syncon the Unexplained seminar was held in one of the prominent spaces just in front of the Wheel proper. After the first Walls Down ritual, this group, which was the second largest, merged with the largest group, the similarly mystical Nature of Man, to form by far the biggest task force, with 39% of the 75 remaining participants.

The history of the Space task force shows the opposite trend. At the May 1972 Syncon, it was designated Task Force A and, with the Environment group, was supposed to "establish the pacing goals" for the world in the decade 1973–1983. It was given prominence as a full-fledged section of the Wheel. The general plan of the later Los Angeles Syncon was the same, but at the May 1973 Washington Syncon, Space was disastrously demoted. The seminar was renamed Extraterrestrial and placed behind the Wheel, far away from the other satellite groups, in a dark corner, screened by a broad partition. Its place in the Wheel was taken by Applied Technology, which might consider some aerospace topics. According to the preconvention plan, at the first Walls Down, Applied Technology was supposed to merge with Environment, Space's partner at the earlier meetings, while Extraterrestrial was supposed to add itself to the merger of Physical Science and Biological Evolution. In fact, the few members of the Extraterrestrial group dispersed, and their official leader was forced to complete its work on his own.

Table 6.2 shows the comparison of participation in the various task forces at the May 1972 Carbondale Syncon and at the one held in Washington almost exactly one year later. The Carbondale figures are from the official registration list, while the Washington figures are based on the first census I performed (see Table 6.1).

There was very little change over the year. The doubling in the size of the Government seminar can well be explained by the fact that the latter Syncon took place not only in the nation's capital, but in the very hotel inhabited by the Vice President in a section of the city congenial to government officials.

The big shift is from Space to Technology. If we add the percentages in these two groups, we get 18.2% for 1972 and 20.4% for 1973, about the same for both Syncons. I believe the decision of the CFF leadership to downgrade Space, communicated indirectly to the membership through the changed Syncon layout and convention advertising, led prospective Space members to switch to Technology. Had Mrs. Hubbard wanted to continue boosting spaceflight, the CFF could have arranged the Syncon contingencies in such a way as to keep the enrollment high in Space; clearly she did not want to do so.

TABLE 6.2 COMPARISON OF TASK FORCES IN 1972 AND 1973 SYNCONS

Task Force Title	*May 1972*	*May 1973*
Space, Extraterrestrial	12.8%	6.8%
Technology, Industry, etc.	5.4%	13.6%
Government	4.7%	9.7%
Social Needs	13.5%	11.7%
Environment	8.8%	11.7%
Other Regions (Third World)	7.4%	4.9%
Information Evolution	8.1%	5.8%
Arts	4.7%	5.8%
Nonverified, Unexplained	15.5%	12.6%
Nature of Man	14.2%	14.6%
Others	4.7%	2.9%
	N = 201 people	103 people

Electronic Mysticism

A very important part of the Syncon process is the continuous closed-circuit television coverage of the various activities and presentations. At the first Syncon in Carbondale, S.I.U.'s radio and TV department provided the necessary equipment. Students manned the cameras and editing machinery and performed interviews. An S.I.U. senior, Robert K. Weiss, operating through his own enterprise called Camelot Productions, coordinated the work and created a number of taped programs under the title New Worlds Video. The students I interviewed who were working on this project seemed quite excited to have the chance to create real television, rather than mere classroom exercises, and presumably the university was pleased for its students to have this experience. The CFF, of course, received much free labor and free use of an expensive TV system.[52]

Leaders of the CFF were talking about buying their own TV system. They hoped to make Syncon a permanent traveling operation and would need to be independent of the facilities of host organizations. Indeed, television would be one of the most attractive features of the package the CFF could offer to any college or other group that might consider inviting the Committee to put on a Syncon.

According to former CFF leader Ira Einhorn, in September 1972 Mrs. Hubbard invested $250,000 in a TV system, which was purchased by Marx family money under the cover of a new corporation, New Worlds Video, Inc., Los Angeles. At this time, Einhorn withdrew from active participation in the Committee because he felt too much emphasis was being placed on technological equipment and too little on planning and finding a clear direction in which to proceed. He also complained that the system was not compatible with standard broadcast TV. Syncon tapes have to be transferred to 16mm film before they can be shown over the air. Einhorn said that Mrs. Hubbard should not have bought so impulsively; she perhaps would have been better advised to wait for a compatible system.

The system is said to consist of 12 black and white cameras, 5 color cameras, 25 color monitors, 4 video tape recorders, auxiliary equipment, and a large van for transporting everything.[53] The Los Angeles Syncon was the first at which this sytem was used, and by the May 1973 meeting it had produced two commercial video tape packages. One was a documentary of a fashion show, made for its promoters, and the other was a short series of training tapes made for a California police department. There is a full-time staff of three, and others are hired for part-time work when there is a special project. Robert Weiss is the executive producer of New Worlds Video.

As a part of the Washington Syncon registration procedure, participants had to sign a television release form, agreeing that the CFF could use TV and audio tapes made at the convention for any noncommercial purpose. Speakers and persons I interviewed repeatedly stressed that condensed tapes of the various Syncons would be broadcast over cable TV systems throughout the country. In advertisements the Committee offers to lend these tapes to interested groups and educational institutions. At the November 1973 Syncon in Huntsville Alabama, the New Worlds Video system was successfully hooked up to the local Cablevision outfit, and proceedings were sent out live to home viewers.

Particularly at the Carbondale 1972 Syncon, TV was used as a communications medium for the participants. Many little reports and presentations were televised and distributed live on monitors in all sections of the Wheel. A special area was also set aside for repeated playing of the video tapes. Anyone who wanted could watch himself and his new friends on the tube, over and over. At the Washington meeting a year later these functions of the TV were far less important. Very few special presentations were televised, and the planned "video sphere" playback area was never actually put in operation.

The CFF says that its video system is "designed to advance synergistic convergence" among Syncon participants and claims it "increases vital interactions and fosters the ability to make decisions based on an awareness of the whole."[54] On the rational technical level, my observations indicate that the system functions very poorly. The only times real information is communicated by Syncon TV is when set speeches are given at intervals reserved for public presentations when the task force work is halted. The participants could just as well gather in a lecture area and hear the speeches directly, rather than sitting in the various sections of the Wheel watching them on the monitors. The TV may, however, serve a number of psychological functions.

Quoting Andy Warhol, a CFF newsletter suggested that through Syncon TV, "everyone will be a star for five minutes." To take advantage of this system for narcissistic display, the Synconner has to adopt the role of a serious world analyst, take the CFF orientation seriously, and execute an adequate performance in playing the Syncon game. Thus, the TV can act as a general commitment mechanism.

At the Carbondale meeting the TV appeared to act as an attitude change device. A number of social psychological studies have demonstrated that, under the right conditions, having persons publically act the role of spokesman for an attitude will effectively move their "real" opinions in the direction of this attitude.[55] At this Syncon a number of participants explained over TV the positive value of the space program. Between live presentations the recorded CBS television coverage of the American space program—all the flights from the first *Mercury to Apollo 11*—appeared on the same screens. The emotional impact of the live programs was perhaps increased by the fact that the participants did indeed act as though they were colleagues of newsman Walter Cronkite, with whose canned image they shared Syncon air time. Now that Syncon has become empty of specific ideological content, other than the general fusion orientation I describe, it of course has lost any function it may have had as a means of changing participants' attitudes in a pro-spaceflight direction.

The CFF communications system is important in a ritual or mythological sense as the concrete manifestation of the nervous system of the emerging fused organism of mankind. The Committee suggested that with the development of New Worlds Video, "a rudimentary self-consciousness and nervous system of a social organism is generated." At Carbondale Mrs. Hubbard attempted to place an international conference-call on the phone, uniting Illinois with France and Yugoslavia. There was a thorough mix-up, with each group unable to hear the others and struggling in fits and starts to recite its prepared ecumenical statement. She explained to the Carbondale audience that this phone hook-up was like the nervous system of a new baby, uncoordinated and erratic, but destined to be a mature unity. New Worlds Video closed-circuit news shows intermingled Syncon reports with world news from the wire services "to help participants recognize that everyone make news daily. There are no observers on planet Earth. We shape history by our every act."

The Carbondale Syncon had a dramatic data retrieval system. Suppose the Space task force wanted to know the annual American production of liquid hydrogen. A member of the group would dash into a cramped communications hut in the center of the Wheel and either telephone the question or write it out for a remote copier to transmit. Students in the university library would receive the question and search madly through the book stacks and reference room. If they found the answer, they would write it out and give it to a courier who would hop on a bicycle and zoom to the ballroom of the student center and deliver the data. This system was not of much use.

The Washington Syncon attempted to tie a time-shared computer into the convention proceedings through a linkage called Syncomp. The CFF claimed Syncomp would be of great use in facilitating interaction between participants,[56] but judging from my observations of people attempting to use the system and considering the paucity and triviality of printout,[57] any practical good of the machinery is hard to see. The system could be used to display messages on selected TV monitors, but most of its business seemed to be the simple storage of group reports in abbreviated form. Some participants were proud to learn how to type in comments and delighted in playing with it or perhaps using it as a tool for sketching out simple ideas. A pad of writing paper would have served just as well.

For members to know that a computer was a part of the proceedings may have been important, because cybernation has become an essential badge of high intellectual fashion, and a futurological conference might have seemed backward without one. Computerese terminology was in fairly common use, often applied inappropriately or awkwardly spoken. One participant had the habit of saying people were "interfacing" when he meant "meeting and communicating." Michael Kernan, writing for *The Washington Post,* noted that "an air of technocracy pervades" Syncon and the CFF's pronouncements, and wrote of the use of language at the meeting:

> At first the various languages of different disciplines clashed as people talked at each other and reached no rapport. However, the language became increasingly generalized, multisyllabic and so inflated with terms like "mankind," "society," and "the universe" that the clashes subsided. People were still talking at each other, but they didn't know it.[58]

Because the major thrust of the Syncon process is toward "synergistic convergence" (consensus, unity, fusion), naturally the cultural content of the meetings becomes both abstract and mystical. As Simmel argues, the most general symbolic representations and the most fluid emotionality are the lowest common denominators that can unite a mass of people into a collectivity.[59] My questionnaire asked Syncon participants what NASA's funding for the next year should be. Sixty-two percent called for an increase, while 38% of those responding called for a decrease. Clearly, if the CFF set space expansion as its chief goal, 38% of the Synconners would dissent. Now that the CFF has taken "a better future" as its goal, probably all agree.

Synconners

Who are the Syncon participants? Let's examine the evidence from the questionnaire, the information provided by the CFF, and my field observations and interviews.

Except for the close-knit CFF contingent itself, Synconners are not the same people from meeting to meeting. The membership lists for the May 1972 and May 1973 Syncons include a total of 411 names, not counting myself. Only 20 names appear on both lists. Synconners tend to be local people brought in by the sponsoring organization, for example, students and faculty of Southern Illinois University and the University of Alabama at Huntsville for the Syncons held at those two schools. A number of others are snared by Mrs. Hubbard's social network or come as representatives of other organizations and causes who plan to exploit the convention for their own purposes.

Like many other social movements, the CFF has attracted a number of "deviant" persons rejected for some reason from more conventional institutions and normal social life, who came to Syncon seeking scope for their creative energies, honor for their talents, happy sociation with peers, and audiences for their self-presentations. The S.I.U. student who designed and built the 1972 Syncon Wheel had been feeling there was little direction to his schoolwork and neither clarity to his future nor social meaning to his labors. He welcomed the Syncon job as "an interesting challenge" for which he was prepared to "quit school." The young woman who was in charge of the information center for Syncon at the S.I.U. library was an artist whose school progress had stalled and who seemed to be suffering a crisis of identity which prompted her to seek value-laden movements to provide both personal direction and social contacts.

Perhaps the most aggressive social mixer at both the Carbondale and Washington meetings was a young crippled man confined to a wheelchair. He was an S.I.U. student who quit school complaining that the special program designed for the handicapped in which he was enrolled was both too restrictive, not permitting him to develop and discover his own desires, and lacking in optimistic encouragement for his personal future.

A display of space paintings was proudly presented at one Syncon by a young man who had just been ejected from the design program at a midwest college because of his deviant attitude toward his studies. In all his courses he refused to draw houses or other prosaic commonplaces, but insisted on sketching great teardrop spaceships complete with spiral staircases and staterooms. At Syncon he found an interested audience that included several persons pleased to discuss the spaceship plans with great seriousness and in great detail.

A multimedia artist about to be "kicked out" of the staid music program at a Massachusetts college charged that his interest in electronic music had turned his teachers against him. Syncon, he said, had changed his entire life. For the first time it had provided him an enthusiastic audience, like-minded people to chat with, and the hope of exciting future projects and experiences.

In general, however, the CFF did not admit such deviants to its inner circle of leaders, but kept them at a distance in the common ranks of participants. For example, one young man who very much wanted to become a full member of the Committee was an editor of the amateur spaceflight magazine *Orion,* which he and a partner had created. He was a dropout from the Air Force Academy and the son of a successful aerospace professional and science fiction author. This young fellow desperately traveled the world trying to assemble a network of space enthusiasts. For a year he tried to work with the CFF, continually offering them his services and sadly being always rejected.

The Committee does seem to need one deviant on its staff, a "token hippie," acting as poet laureate. When the first heavily bearded, free-verse writer left the group, a second was immediately appointed in his place. This institutionalized role appears to express the group's concern that all kinds of people in the world come together and work cooperatively for a positive future.

Many Synconners, however, are "solid citizens." Several claim ranks or titles. The 411 on the two membership lists include two Reverends, a Major, a Captain, an M.D., two Professors, one Ambassador, and 28 Dr.s. Of the 80 who responded to my questionnaire, only 2.6% had a high school education or less; 24% had received some college education; and an equal number had finished college without undertaking graduate work. A Master's degree or the equivalent was held by 32.1%; 16.7% claimed Ph.D. degrees.

Students accounted for 21.9%; 12.3% worked for educational institutions; 13.7% had jobs in government, 17.8% in businesses, and 2.7% in private foundations. There was extreme variation in reported incomes. The mean was $12,500; the standard deviation was $12,400.

The mean age of 35.1 years and the standard deviation of 13.9 reflected the fact that the group consists mainly of young and early-middle-aged adults, but contains people of a wide range of ages. Twenty-one and eight-tenths percent were women; 45.9% were married, and because those who were married tended to have a couple of children, there was exactly one child reported per respondent.

Religions included 24.7% Protestants, 16.4% Catholics, 8.2% Jewish, and 20.5% "nonreligious." A large number, 30.1%, chose the category "other." This undoubtedly reflects the fact that the CFF has become a meeting group for a number of occult and mystical groups. I included in Q-CFF and in the questionnaire sent to members of the New England Science Fiction Association in 1973 a question on extrasensory perception which the British science news magazine, *New Scientist,* had already given its readers. In Table 6.3 I present the results.[60]

We can see that the science fiction fans of NESFA responded in a way very similar to that of the readership of *New Scientist,* while Syncon participants included a much larger number convinced that extrasensory perception is an established fact. Within the CFF sample there was a very interesting subset of 11 people who believed that contact with beings from another planet had already been made. This group overwhelmingly considered ESP to be an established fact (80% compared with 51.9% for the whole sample) and were much younger, with a mean age of 25 compared to 35.1 for the total sample. This subset did not represent any one occult organization.

I also presented respondents with a question taken from a World

TABLE 6.3 BELIEFS ABOUT ESP

	Established Fact	*Likely Possibility*	*Remote Possibility*	*Impossibility*	*Unknown*
New Scientist	25%	42%	19%	3%	12%
CFF	51.9%	33.8%	7.8%	0%	6.5%
NESFA	17.1%	48.6%	24.3%	0%	10.0%

Future Society questionnaire, which asked about the quality of life we will experience in the year 2000, compared to today.[61] Comparisons of the three groups of respondents are shown in Table 6.4. The science fiction fans and world futurists respond similarly, while the CFF Synconners seem unusally optimistic.

The World Future Society often sends official representation to CFF Syncons and several Synconners say they are also members of the Society. In addition, a number of small futuristic think-tank organizations have sent representatives to Syncons, including The Institute, Turtle Bay Institute, Thomas Jefferson Research Center, and Futuremics, Inc. Not only the WFS's magazine *The Futurist,* but also pop-futurology publications *Futures Conditional* and *Synergy Access* have been distributed at gatherings.

Aerospace companies and agencies usually sent a few representatives to Syncons, propagandizing for space in general or the company's projects in particular. The May 1972 Syncon was held just before the federal government decided to award the chief contract for the Space Shuttle to North American Rockwell, and both this firm and its rival McDonnell-Douglas sent representatives to the convention. Sales literature from NASA and the Department of the Interior propagandizing for the ERTS earth resources satellite system was distributed at the 1973 Washington Syncon.

The aerospace companies come to Syncons as valuable but peripheral guests, while at least three occult groups have gained significant influence within the CFF: the Prosperos, the Cooperators, and AUM. In addition, a variety of other farout organizations have sent representatives to Syncons. For example, Wayne Sulo Aho of the New Age Foundation, who claims to have been picked up by a flying saucer on May 11, 1957, was pleased to render "an assist" to the Extraterrestrial task force at the D.C. Syncon.

TABLE 6.4 PROJECTIONS ABOUT QUALITY OF LIFE IN THE YEAR 2000

	Much Better	*Somewhat Better*	*The Same*	*Somewhat Worse*	*Much Worse*
WFS	10.8%	56.3%	11.4%	19.6%	1.9%
CFF	44.1%	30.9%	7.4%	10.3%	7.4%
NESFA	11.8%	51.5%	8.8%	23.5%	4.4%

Thane Walker, "dean" and founder of the Prosperos of Santa Monica, California, met Barbara Hubbard quite by accident in a Washington hotel elevator in 1970. He attended the Carbondale Syncon, calling it "one of the highlights of my entire life." Another Prosperos member told me, "He went to Carbondale, and he came back and explained it to us. And we knew it was just in the same vein we were working in, so we were all interested." This informant estimated that 45 or 50 Prosperos students went with Thane to the Los Angeles Syncon, where he was moderator of the Nature of Man seminar. At Washington Thane had the same job, but perhaps because of the traveling distance, only between six and 10 others came with him. The director of New Worlds Video is a member of the group, and considerable publicity for the CFF has been given in the Prosperos newsletter *Trends.* In announcing that Barbara and Earl Hubbard would receive the Prosperos Award of Merit for 1972, Thane observed: "[T]hey, like we, believe that mankind's future lies not in concepts and identities of the past, but in the new identity of man as native of Eternal and Universal Space."[62]

The Prosperos teaches its own occult doctrine, and if the CFF can be seen as a deviant or counter think-tank, the Prosperos, like many other occult groups today, can be seen as a deviant or counter university. A Prospero told me:

> I dropped out of college about four years ago because they weren't teaching me what I wanted to learn. I wanted to learn about me, y' know....I went through Scientology, a lot of Eastern things....I went through drugs....I hold a degree right now which is the equivalent to a Bachelor's degree. It's called The High Watch—the inner body of the school.

Teachers in The Prosperos append the initials *H.W., M.* after their names, standing for the Thane-conferred degrees of High Watch and Mentor, and teach such deviant subjects as "Science Fiction," "Releasing the Hidden Splendour," "Transcendental Sexuality," and a metaphysical teaching called "Translation." The Prosperos advertise in the central occult monthly magazine *Fate.* The doctrine of the group includes a number of Teilhardian notions and, like the CFF, urges personal evolution toward a universal species. In an open letter, Thane announces:

> With the startling discovery of the "atomic" nature of all existence and the leap into space, mankind today is on the threshold of the greatest transformation since creation. Whether he likes it or not, *a new identity is being forced upon man.* But man must *drop off* the old "flesh and blood" identity to experience the astounding revelation of a new Transcendent Self that is *master of space and time.*

Like Oberth and the CFF, the Prosperos tie ontogeny to phylogeny and go so far as to advertise "Cosmic Intention Therapy," "basic to other classes," designed to promote individual evolution and based on the following principles, according to a course booklet:

> The Life Force that has transcended itself in every form of life as we have known it has now produced a new creature—the spaceman. Mankind as a whole must make this evolutionary leap intelligently, preserving the best, the noblest, and the highest concepts of earthman's culture, and this can only happen at the individual level.

The Prosperos is one of more than 50 organizations that are members of the ecumenical International Cooperation Council in which Irving Laucks, who pledged the $100,000 to endow the Unexplained Phenomena Syncon task force, is a major influence. Laucks is founder of a syncretic religious group called the Cooperators which publishes a bimonthly magazine, *The Cooperator,* jointly with the International Cooperation Council. Like the CFF and the Prosperos, Laucks' group advocates a view of man that places him in control of his own evolution and gives him the duty of "changing human nature" and uniting in pursuit of a greater destiny in the universe. Laucks calls the purposeful control of the future "the Third Evolution" and says in a brochure distributed at Syncon that this concept

> ...can meet the requirements of an ecumenical religion for Earthmen, of an alliance of science and religion, of offering to man hope of the prospect of being a worthwhile and an important member of an interesting future existence and society, far surpassing anything he has ever experienced in this earthly life. For there is no end to evolution, and no end to the interesting possibilities it offers.

At the May 1973 Syncon I noticed that Barbara Hubbard frequently sought the advice of Bob Hieronimus, coordinator of the Unexplained Phenomena section and founder of an esoteric school in Baltimore named AUM. According to literature he distributed, AUM offers a large number of courses on occult topics such as "Astrology," "Yoga," "Kabballa," "Tarot," and "Visual Alchemy" and has as its purpose "sowing the seeds of the Aquarian Age," which "symbolizes altruism, brotherhood, synthesis and unity." Not only does AUM's ideology resemble that of Syncon, but Hieronimus told me that AUM "in many ways uses the same techniques as Syncon." Similarly, not only does the Prosperos' creed resemble that of the CFF, but according to one of its leaders, this group has actually put on its own Syncon meetings.

All these occult groups seem to prefer continual renewal of their plans, running through a series of short or aborted projects at great speed. While this would not be an appropriate attitude for a technical aerospace organization, which would be best advised to carry long major projects steadfastly through to completion, the almost whimsical, impulsive style of these groups serves to entertain and excite their memberships. It keeps the movements going as social enterprises, even if the explicit tasks they tackle are frequently abandoned soon after they are first taken up.

Instability and the Quest for Meaning

The May 1972 Syncon ended with the dramatic announcement that an on-going Syncon would be set up in Philadelphia. Indeed, a continuing organization with intermittent Syncons, if not exactly a continuous "synergistic convergence," was established, but the Philadelphia base of operations was maintained for only one year. Similarly, the May 1973 Syncon was also capped with the dramatic announcement that the *Community* for the Future had been founded, an ambitious project intended to "synergize efforts of people and groups so as to develop a world-wide Movement for the Future."[63] This announcement represented a rebirth of the CFF and explained that the Syncon marked the end of "Phase I" of the movement and the beginning of "Phase II." The idea was to bring more people directly into the standard activities of the CFF and to institutionalize support.

There were to be six official grades of membership distinguished by the size of the financial contribution made. Barbara Hubbard's toy magnate father, Louis Marx, became the first founding member in February 1973, with the promise of a $100,000 donation. A few months later the CFF sent letters to those who had joined the Community, announcing that the "basic membership plan" had been completely changed so that there would be only one grade, with a standard $10 a year dues. This indicated a scaling back of the whole idea to a level of activity indistinguishable from "Phase I." The one major symbolic change was that one could become a bona fide *member* of the CFF, rather than merely participating in individual CFF projects. However, even the membership card promised in this letter was soon forgotten and never produced.

The Committee for the Future has no formal decision structure. There is a board of directors, but Barbara Hubbard is Chairman and Chief Executive Officer, and the real authority rests in her hands. She is of course strongly influenced by Whiteside and others, but her whim rules. Had some more formal executive structure been developed, greater stability and some responsiveness to the general membership might have resulted. Lacking the governor of a formal constitution or vested aerospace interests, the inertia of a body politic, or any other force to balance Mrs. Hubbard's willingness to shift focus frequently, the CFF follows a kind of "drunkard's walk" through a wide territory of potential ideological and activist commitment. Bound to spaceflight only by the personality of Earl Hubbard, it may be bound to Syncon only by that of John Whiteside. The only prediction we can make about the future of the CFF is that there will be a sequence of changes and shifts in purpose.

The statements Mrs. Hubbard has made about her own intense quest for meaning are classical expressions of *anomie*—which is usually defined as the state of normlessness, being without effective rules for living or standards of self-judgment. In strong language of which the CFF might approve, Peter Berger describes the state of humankind as one of great psychic danger. "We are surrounded by darkness on all sides as we rush through our brief span of life toward inevitable death."[64] In her search for "glory untold," Mrs. Hubbard was convinced by the idealistic rhetoric of her husband to believe that transcendence of the human condition could somehow be achieved through spaceflight.

The cure for anomie is to be found in an intense involvement in society, in any human pursuit carried on in the fellowship of other persons. Writing with Thomas Luckmann, Berger says that "the institutional order represents a shield against terror. To be anomic, therefore, means to be deprived of this shield and to be exposed, alone, to the onslaught of nightmare."[65] In another place Berger speaks of anomie as "radical separation from the social world," and says it represents "a powerful threat to the individual" leading to a conditon in which he "loses his orientation in experience."[66]

When Barbara married Earl she presumably submerged her anomie in the meaningfulness of his world. Existentialist psychoanalyst Viktor E. Frankl argues in his book *Man's Search for Meaning* that "The salvation of man is through love and in love."[67] If anomie is separation from a meaningful world and meaning is a social construction, successful love is an antidote to anomie. Sociologist Pitirim A. Sorokin tells us:

> [*L*]*ove is the experience that annuls our individual loneliness;* fills the emptiness of our isolation with the richest value; breaks and transcends the narrow walls of our little egos; makes us coparticipants in the highest life of humanity and in the whole cosmos; expands our true individuality to the immeasurable boundaries of the universe.[68]

Mrs. Hubbard's anomie was compensated by the meaningfulness of her marriage, at first perhaps in the excitement of starting an intense relationship, later simply by the day-to-day business of taking care of a husband and five children. As Kipling noted in "The Supports," a poem on the simple things in life that keep us going:

> *Heart may fail, and Strength outwear, and Purpose turn to Loathing,*
> *But the everyday affair of business, meals and clothing,*
> *Builds a bulkhead 'twixt Despair and the Edge of Nothing.*

The reemergence of her anomie coincides with the maturity of her children. When the business of motherhood is done, what are mothers to do? Perhaps, sometimes, they can give birth to a different kind of child, like the Committee for the Future. Barbara's children were grown or in the custody of a housekeeper when she threw herself into spaceflight boosting. But the conquest of the universe is itself an anomic goal. Durkheim long ago noted that anomie can result from the loss of any ceiling to our aspirations:

> All man's pleasure in acting, moving and exerting himself implies the sense that his efforts are not in vain and that by walking he has advanced. However, one does not advance when one walks toward no goal, or—

> which is the same thing—when his goal is infinity....To pursue a goal which is by definition unattainable is to condemn oneself to a state of perpetual unhappiness. Of course, man may hope contrary to all reason, and hope has its pleasures even when unreasonable. It may sustain him for a time; but it cannot survive the repeated disappointments of experience indefinitely.[69]

The premature attempt to colonize the solar system could sustain Mrs. Hubbard for a time, particularly because it involved intense social activity, but as soon as Syncon presented itself as an alternative, it became the center of her hopes. Syncon, after all, was in essence pure social interaction. The manifest agenda may have been world planning, but the hidden agenda was love and the fusion of the individual into the group, what Berger and Durkheim might have called religion.

We can compare Mrs. Hubbard's instability of commitment with the stability of the early rocket pioneers. In Chapter 2 we noted the almost mystical orientation of the three great theorists. Tsiolkovsky and Goddard had *peak experiences* (perhaps we should call them *religious conversion experiences*) at the time of their dedication to spaceflight. Oberth had an intense interest in occult ideas throughout his life. In this they were like Barbara Hubbard. But the rocket pioneers were able to go immediately on to realistic, successful, personally gratifying *work* on behalf of their dreams. They experienced joy each time a calculation or an experiment took them one step nearer to their goal. Indeed, for all the early space engineers the "great leap" to the moon was not an anomic gulf, but a road of many "small steps" each one a goal in itself.

In Chapter 5 we saw that all the founding members fo the AIS and BIS dropped out in the first few years, except one man in each, Pendray and Cleator. The initially fanciful membership of both groups was replaced with practical men able to gain day-to-day satisfaction by performing calculations and experiments. Pendray and Cleator are precisely exceptions that prove the rule. Both served as historians for their rocket societies, organizers, and finally pioneers emeriti. They played special roles and received special satisfactions.

Neither Mrs. Hubbard nor any of her lieutenants were able to find a source of continuing satisfaction in space boosting after NASA had turned a cold shoulder on their romantic projects. She was hardly prepared to design starships or take any other practical but modest action that would technically further planetary colonization. Social satisfactions came from the CFF conventions, which could better be justified as world planning sessions.

A good test of the CFF's thorough disaffection with space was provided by the November 1973 Syncon held in Huntsville, Alabama, "the missile and space capital of the universe."[70] Huntsville is the home of the U.S. Army Missile Command, NASA's Marshall Space Flight Center, and most of von Braun's team of German-American rocket engineers. If the spark of spaceflight interest was ever going to be rekindled in the CFF, it would happen in Huntsville.

But there was no Renaissance of space enthusiasm. The local sponsors certainly wanted to use the Syncon to advance the cause of spaceflight. The four men who arranged for the meeting were James Beal and Konrad Dannenberg of NASA and Donald Tarter and Don Smith of the Sociology Department of the University of Alabama at Huntsville. The two sociologists were both strong supporters of the Spaceflight Movement, and Dannenberg had been a member of von Braun's Peenemünde team. He had been head of the German *Taifun* antiaircraft rocket project and executive of the Jupiter IRBM team at the American Army's Redstone Arsenal. He is currently active in an organization called the International Association of Educators for World Peace. One of the four told me NASA had little respect for the CFF, but despite his own misgivings about the Committee felt Syncon was worth a try.

The title and focus of the meeting was "Technology's Impact on Society," and the advance CFF literature did not emphasize space, except in implying that a number of *important* NASA people would grace the affair with their presence. Although some CFF low-status members have suggested that the Syncon Wheel be adjusted to the needs of each different conference, Whiteside's standard plan was followed in the Washington form, with Extraterrestrial set up as a peripheral seminar.

The final published summaries of the panel discussions mention space slightly more prominently than those from the Washington Syncon, but the primary emphasis is on world unity and development, communication, and the energy and ecology crises. In the intermediate, Some Walls Down stage of the Syncon, Extraterrestrial merged with four other panels, and there was a bloc of five social issue panels. Unexplained Phenomena and Nature of Persons (the new nonsexist name of Nature of Man), which were the largest groups at the other meetings, formed a two-panel bloc. Again the occult-philosophical groups were strongest, and Extraterrestrial was weak.

According to the official membership list, 52.4% of the 168 participants were Huntsville residents, and another 6.5% lived elsewhere in Alabama. Discounting the 8.9% CFF official contingent, only 32.1% came on their own to the conference from other states. Indeed, I recognize the names of several CFF regulars in this last category, and there was a strong group from the University of Tennessee, not too far away, where the next Syncon was to be held.

This membership list also gave very brief occupational descriptions for the participants, and in Table 6.5 I present the result of tabulating these data by categories. The general picture of the conference is of a university-based extracurricular activity combined with a traveling carnival of CFF regulars.

CFF literature mailed after the meeting shows no shift in the direction of renewed interest in space at all. Except for an uncritical and understandable flurry of excitement over the great significance of the Syncon, which necessitated some minimal mention of space but stressed the TV set-up much more strongly, the emphasis was soon on the next Syncon and on ballyhooing the New Worlds Center in Washington, a small countercollege.

The history of the Committee for the Future is interesting in two

TABLE 6.5 HUNTSVILLE SYNCON PARTICIPANTS

Occupational Categories	*Number*	*Percent*
Committee for the Future	15	8.9%
World Future Society	5	3.0%
NASA, etc.	7	4.2%
Students	59	35.1%
Professors and teachers	22	13.1%
Press	2	1.2%
Companies, government agencies	18	10.7%
Military	2	1.2%
Various institutes	3	1.8%
No data (some unemployed)	32	19.0%
Self-Employed	1	0.6%
Hypnotists	2	1.2%
Total	168	100.0%

ways. It suggests that there is little or no opportunity for amateurs to participate in furthering the exploration, exploitation, and colonization of the solar system, and that the Spaceflight Movement has reached a level of maturity that makes room for no one but qualified professionals in its ranks. But the experience of the CFF also suggests something about the way commitments are created and maintained in radical scientific and technical movements. The initial urges that brought both the CFF and the Spaceflight Movement into existence were noneconomic, impractical, personal, and primitive desires. They could be described either in psychiatric or religious terms. For the earliest space pioneers these archaic urges could be translated into effective social action and workable technical developments. For the members of the CFF this was impossible, and those urges then expressed themselves more directly as religiously colored social concerns and an occult mysticism of joy.

Chapter 7 / THE SCIENCE FICTION SUBCULTURE

Science fiction is the popular culture of the Spaceflight Movement. For more than a century science fiction stories about spaceflight have been widely read, and as we saw in Chapter 2, early SF authors had a direct and significant influence on the rocket pioneers and on the development of their ideas. If a mass popular following for the Spaceflight Movement exists, we should expect to find it among the most avid readers of science fiction. In this chapter I describe the science fiction subculture and the tight network of fans known as *fandom.* Almost every analysis leads to the same conclusion: Although SF may have a variety of social functions and some power to spawn new and deviant movements, it no longer has any direct relationship to astronautics and may be left without a role to play in the further exploration of space.

Origins of the SF Subculture

On April 5, 1926 the inaugural issue of *Amazing Stories,* the first true science fiction magazine, appeared on the news stands, and within a few years a genre had been created, complete with neologisms, literary norms, professional specialty authors, dedicated readers, and fan clubs.[1] The editor of *Amazing Stories,* Hugo Gernsback, did much to stimulate the emergence of the network of fans and clubs—fandom—by publishing letters to the editor complete with the writer's address. Other magazines took up this practice, and fans were able to discover each other through the letter columns and begin correspondence that led to the formation of clubs.[2] Many current professional authors were first mere fans and members of these clubs, as have been some of the most influential editors; so fandom is well integrated with the professional production of SF. The total phenomenon takes place within a single tight social network.

Arthur C. Clarke has compared Gernsback favorably with Jules Verne,[3] and he is seen by fans as "the father of SF." The major awards given out at the annual World Science Fiction Convention, the *Oscars* of the field, are called *Hugos*. There is a Gernsback Crater on the Moon. For most of this century, Gernsback has been a successful and prominent editor and publisher of electronics magazines and how-to-do-it electronics books, and the propagandizing fervor he injected into science fiction was a general boosting of technological progress rather than a specific advocacy of spaceflight. We might view Gernsback's influence as negative if we were only interested in the effect of SF on the general public. According to critic and author Damon Knight, "...science fiction has gone through more than half a century of concentrated development underground, in the ghetto world created by publishers like Hugo Gernsback."[4]

SF became "ghettoized" around 1930. It split off from and became mutually antagonistic with "mainstream" literature. It also became divorced from the actual development of rocketry to develop as an independent and possibly irrelevant subculture.

To chart the development of the science fiction field we must find at least one good index of its strength. The consistent center of SF has been the magazines until the past five years when paperback anthologies have taken over some of the functions of the periodicals. *Amazing Stories,* 50 years old, is still being published, and the most influential and popular SF magazine of the last three decades, *Analog,* first appeared in 1930 under the title *Astounding Stories.*[5] Despite changes in style, the general format of these magazines has remained remarkably stable over this entire period. They contain science fiction and fantasy stories, episodes of serialized novels, factual articles, editorials, book reviews, announcements of fan events sometimes, and often letter columns.

Perhaps the best index would be sales or subscription figures for the most popular magazines. Unfortunately, until recently subscription figures were kept secret, and those estimates that were made public are open to suspicion. The annual almanac of periodical publishing is *The Ayer Directory,* designed primarily for use by advertisers.[6] Some of its volumes contain various estimates of the circulation of a few SF magazines. *Amazing Stories* claimed a circulation of 104,117 in 1928, and 49,383 in 1940 when it had already lost its leadership in the field. *Astounding Science Fiction,* later called *Analog,* took the place vacated by *Amazing Stories.* In 1952 *Astounding* was at a peak circulation of 105,700, dropped to 89,153 in 1957, and was down further to 77,124 in 1962 when it was surpassed by *Galaxy* which was then claiming around 95,000. As *Analog,* it gradually regained strength, achieving 99,228 in 1969 and 116,521 in 1973. These data do not permit much analysis, but the peak figures of 1928, 1952, 1962, and 1973 are all around 100,000. There is no sign of a jump in sales around the time of the first Apollo moon flights in the more detailed recent figures. Although the data on circulation are very poor, we have excellent data on the different issues of magazines and stories published in each year.

In 1952 Donald B. Day published his definitive *Index to the Science-Fiction Magazines 1926–1950,* which lists each story and article published in any English-language SF magazine. In 1966 Erwin S. Strauss and the MIT Science Fiction Society published a supplement to Day's volume bringing the index up to date through 1965. Since then the New England Science Fiction Association (NESFA) has published still further volumes completing the work through 1973. I refer to these works collectively as *The Index,* but they use somewhat different formats and do not code entries in exactly the same way, so care has been required in using them as a unified source.[7]

Using *The Index,* I counted the number of issues of SF magazines published each year, and by sampling the pages of story listings, estimated the number of stories published in those issues. I estimate that the total number of items of all kinds published in SF magazines was 8,000 between 1926 and 1950 and 14,900 from 1951 through 1965. By dividing the number of stories for each period by the number of issues, we get the average number of stories per issue. In table 7.1 we see the results.

The most striking feature of Table 7.1 is the gradual rise of magazine production to a peak in the early-1950s and a decline thereafter. The estimated number of stories per issue rises in the beginning as the magazines develop standard formats, then holds steady around 7.0, except for a big drop to 5.0 right after World War II. The physical size of the magazines was temporarily reduced during this period, primarily because of increased paper costs, so the reduced number of stories per issue is probably the result of this economic factor.

TABLE 7.1 PUBLICATION OF SCIENCE FICTION MAGAZINES AND STORIES

Five-Year Period	*SF Magazine Issues*	*Estimated Stories, etc*	*Stories per Issue*
1926–1930	127	624	4.9
1931–1935	184	1088	5.9
1936–1940	257	1784	6.9
1941–1945	334	2472	7.4
1946–1950	403	2032	5.0
1951–1955	910	6370	7.0
1956–1960	767	5175	6.8
1961–1965	482	3355	7.0

In Illustration 7.1 I graph the magazine issues year by year. We can see in greater detail the fluctuations over time. To supplement this "issues index," I also constructed an "authors index." Using *The Index,* a file card was filled out for each author listed in the alphabetical author

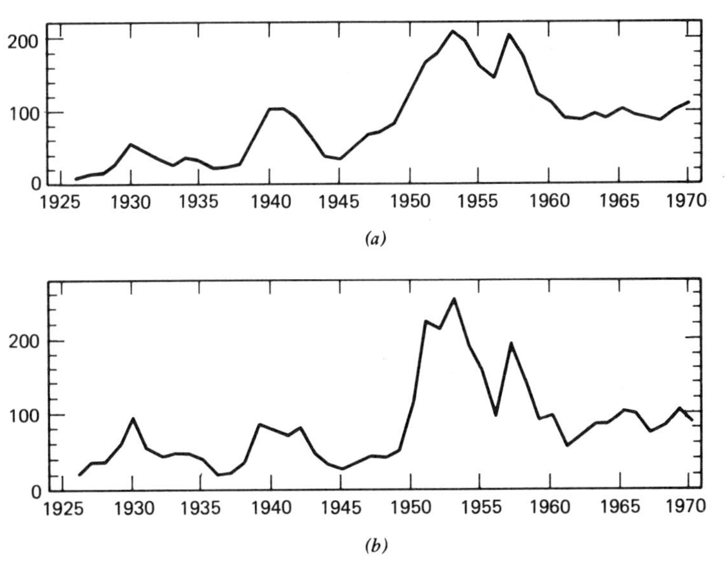

Illustration 7.1 Science fiction, 1926-1970. (a) *Number of issues of English language science fiction magazines published each year.* (b) *Number of new author names that first appeared in science fiction magazines each year.*

section, giving the year in which that author first appeared in the science fiction magazines. The cards were then sorted by year and counted. The chart shows the number of new author names per year of first appearance. Although the list is exhaustive, including 4058 names, a major problem is that we have no way of purging the list of pseudonyms. This does not matter for authors who write only under a single pen name; for example, Harry Stubbs is the real name of a science teacher who writes entirely under the name, Hal Clement. Sometimes a pair of authors will collaborate in writing stories under a third name; for example, Henry Kuttner and his wife C. L. Moore collaborated in producing the works of Lewis Padgett. Frequently one author will use several pseudonyms, perhaps writing his best work under his own name and throwing away pen names with poor short stories. An author not only protects his reputation with pen names but also increases the number of stories he can sell, avoiding becoming a glut on the market. In the early-1940s Heinlein was forced to use several names for just this reason. Of the 1403 complete author references in *Day's Index,* 185 (13.2%) are known to be pseudonyms. We can hope that the ratio of names to real authors is roughly constant or at least changes slowly with time, and we use this author index as a reasonable supplement to the issues index.

As Illustration 7.1 shows, the two curves are extremely similar. I would argue that each measures one kind of activity in the field directly, and that both are more than mere substitutes for the missing sales data. Because the *Ayer Directory* presents estimates for the total number of periodicals of all kinds published in each year, it struck me as a good idea to compare some of these data with our issues index. Because SF magazines usually aim at monthly publication, I collected figures for the total number of different monthlies published in the United States each year, as reported by the following year's *Ayer Directory*. The correlation between these figures and the number of issues of SF magazines published for the 45-year span 1926–1970 is +0.46, showing a moderate relationship. *Ayer's* figures include some but not all of the SF magazines, and their number is tiny compared to the total of all monthlies, so I felt that correcting for the inclusion of SF in the Ayer figures was not necessary. The correlation might be explained by the fact that both variables trend upward over time. The SF indexes are much more volatile than the figures from *Ayer*. The issues index has a coefficient of variation of 0.62; that for authors is 0.65; general monthly magazines show a coefficient of only 0.11.

With some care, we can interpret these curves. There is a rise at the beginning, when SF gets off the ground. The 1930s show first a decline, then a leveling out. There is some disagreement about the impact of the Great Depression on the SF magazines. Alva Rogers contends that SF magazines and pulp literature as a whole, served as a means of escape from the troubles of the day for a large number of people, and that therefore "the Depression had little effect."[8] However, the parent companies who published these magazines were often hard hit, and even the most popular titles sometimes had to suspend publication temporarily.

There is some evidence that the rise in the late 1930s was the result of structural changes in the market. Science fiction and fantasy magazines were merely one segment of the popular "pulp" fiction periodicals.

> The brief 1937–38 recession experienced by the country managed, paradoxically enough, to stimulate fantasy publishing. Because even such standbys as detective and western story magazines showed slumps in sales, pulp chain executives were more than willing to investigate any medium showing possibilities of profit.[9]

A number of these chains came out with their own SF magazines, and "an orgy of science fiction publishing...was under way."[10] We would expect some decline to follow, merely as a reaction to overinvestment in the field, as competition drove some of the new periodicals out of the field.

The valley in the 1940s comes soon after the United States entered the war and is probably a response to scarcities and economic changes brought on by the war. Science fiction was but one of several kinds of popular fiction published in "pulp magazines." According to Theodore Peterson's history of twentieth century magazines:

> During and immediately after World War II, publishers of pulps were hit especially hard by swiftly rising production costs, which increased 72 per cent between the end of 1944 and the middle of 1947. Their revenue from circulation was no longer ample to support them.[11]

The Big Boom of the Early 1950s

The most prominent feature of our charts is the mountain of publications which peaks in 1953. At the time fans were conscious of a boom in their favorite literature, and some imagined that the rise would continue until SF took over the field of publishing. The "Big Boom" carried with it propaganda for the Spaceflight Movement, but also a variety of special SF values such as imagination, intellectual unconventionality, the desire for a grand future, and the hope to remake the world in the name of science. SF fans hoped the boom would install them and their values nearer to the center of American culture. The "Big Boom" is recalled with nostalgia and the sense that something really important happened to science fiction in those years.

> The fifties saw the Big Boom in science fiction during which an incredible number of science fiction magazines came on the scene, and it also saw, by the end of the decade, the Big Bust which reduced the field to a handful of hardy survivors.[12]
>
> The great boom in science-fiction magazines that started in 1949, reaching its peak in 1953, provided a golden opportunity for new talent. A score of young science-fiction writers found they could sell almost anything they wrote as fast as they could write it. Some of them had little difficulty in selling forty or fifty stories a year.[13]

Science fiction and the little world of fandom were the subjects of a major article in *Life Magazine,* May 21, 1951. Later advertisements for a book club quoted *Life* as saying there were 2,000,000 science fiction readers in America. Actually, author Winthrop Sargent had merely guessed there might be 20,000 hard core fans and a total of perhaps 2,000,000 who read some SF. He said SF was then "one of the most spectacularly booming departments of the publishing business."[14] Like many other commentators at the time, he attributed the rise in interest to the impact of real scientific breakthroughs, calling SF "an infection that had been spreading in this country since the atom bomb fell on Hiroshima."[15]

Business Week headlined an article: "Science Fiction Rockets Into Big Time" and reported that SF was making a lot of money for a few movie producers and might be a good investment for publishers as well. It also attributed the boom to increased public awareness of developments in science and technology: "...in 1945, the atom bomb fell, and science fiction exploded with it."[16] *Publishers' Weekly* announced "Progress in Science-Fiction: No Boom, but a Solid Market." Both journals agreed that the avid SF fan read everything he could get his hands on, so the market was able to grow better in quantity of titles than in copies of each book or magazine. Although one anthology had sold 33,800 copies, "The nucleus of devoted science-fiction fans and collectors may still number no more than 2-3,000 buyers per title...."[17]

While several commentators mention the atom bomb, SF author and editor Judith Merril believed the boom was stimulated by a general growth of interest in spaceflight.[18] The first American missiles were being designed, flights of the V-2 were publicly reported around 1949, and a number of popular books and articles were being published by leaders of the Movement, such as Willy Ley and Wernher von Braun. That the Spaceflight Movement stimulated science fiction sales is certainly plausible, but if this is true, SF had already become the passive partner, depending for its expansion on public enthusiasm roused by real science, rather than playing a decisive active role in furthering the cause of spaceflight. Evidence in support of this view can be seen in the spike in both curves at 1957. As the figures for 1956 show, the two indexes had almost dropped to the constant equilibrium level of 1960–1970 before this brief leap. 1957 was the year of *Sputnik I,* but the satellite was not launched until October 4, so the peak is in anticipation of the first launchings. Although *Sputnik* was a surprise, the imminent launch of American satellites had been widely publicized.

The 1957 peak was probably caused by publishers' expectations of public interest, which had been essentially discounted by the time the actual satellites achieved orbit. Extremely wide publicity was given to the American Vanguard project throughout 1956 and 1957. *Satellite Stories, Venture Science Fiction,* and *Science Fiction Adventures* published only in this period, and in 1958 there appeared the single issue of *Vanguard Science Fiction.* Considering how quickly the index dropped again, we can surmise that the 1957 peak was indeed caused by publishers' hopes rather than by readers' demands.

The constant level in publishing after 1959 or 1960 is also of great interest. The 1960s were the period of spectacular success in real space missions. There is no evidence of a science fiction boom around the time of the first actual landings of men on the moon. The developments of science fiction and of real spaceflight no longer run in step with each other; indeed, SF can be described as greatly estranged from the effective elements of the Spaceflight Movement of which it was originally an essential part. The failure of the curves to respond to real spaceflight history after 1958 suggests that except occasionally in the minds of some publishers SF popularity cannot be quickly explained as the response to the atom bomb, rocket development, or other real science events. What decade in the last century has not contained a spectacular technological development that might have been blamed for any rise in SF that happened to follow it? Perhaps the big boom was not caused by increased public interest in science and technology stimulated by the atom bomb or rocket progress.

In a book on science fiction movies, John Baxter claimed that the commerical success of SF films in the early 1950s was "reflected in" the growth of SF publishing.[19] If Baxter means to say that the big boom in magazines and books was caused by a boom in movies, he is wrong. The appendix to his book contains a listing of science fiction movies along with the dates they were issued. I tabulated these dates and discovered that the peak year was 1957, in which 11 on Baxter's list were released. To get a bigger sample, I turned to the book *TV Movies* edited by Leonard Maltin.[20] This book lists and briefly describes about 8000 movies released to television by 1969. I went through the book and tabulated the release dates for all American films that appeared to be science fiction by their descriptions. For this sample the peak is 1958 in which 21 appeared, with 1957 coming in a close second with 20. Maltin's listings made it possible to correct for general fluctuations in movie production. I sampled the book to arrive at an index of overall movie production, divided this into the list of SF films, and found that the peak year was now 1957, the year of *Sputnik I,* not the year of the great SF magazine boom which reached its high in 1953. Differences between SF magazine and movie production for the decade of the 1950s are not all that great, however. Looking at the 12 years 1949–1960 I calculated the mean year of appearance for the four indexes. That is, I summed the year of appearance for all items and divided by the number of items. For new authors the result was 1954.24; for magazine issues, 1954.59; for films in Baxter's list, 1954.70; and for films in Maltin's list, 1955.81. Because the average of 1949 and 1960 is 1954.5, we can see that the figures are not only close together, but cluster on the middle of the decade. Whatever the data prove, they do not support Baxter's theory that the boom in films was the driving force.

An alternate possibility is that the great boom was entirely an artifact of conditions in the publishing industry. It comes at the time that television sets proliferated across the country. By 1955, 67% of all American households had at least one TV set.[21] The growth began and had its most immediate impact around the larger cities which had always been the major markets for science fiction and perhaps also for other kinds of pulp literature. Adventure stories of many kinds had been popular in pulp magazines for decades.[22] There were pulp magazines dedicated to Wild West stories, detective stories, jungle stories, aviation stories, and general adventure. Radio dramas had been competing with the pulps for 20 years, but television was the force that undoubtedly killed off the last of them. A few magazines were able to survive by changing their formats; comic books apparently found a wide market; and the SF magazines were untouched. Science fiction has seldom fared well on television, and almost all SF programs of the 1950s were aimed at children. Pulp publishers saw many of their magazines dying and naturally attempted to stay in business by intensifying their efforts in those areas that were surviving television. According to this hypothesis, we should be able to find a similar boom in other genres of pulp fiction that survived the 1950s.

The obvious example is comic books, and appropriate data can be found in Overstreet's *The Comic Book Price Guide,* a catalog like those long published for stamp and coin collectors, giving suggested prices for antique comic books.[23] The 1974 edition of *Overstreet's Guide* contains 414 pages of listings. I used all material on 94 appropriately selected pages (23%) as my sample. On these pages 333 titles were listed for comics published only in a single year, often only a single issue. These were ignored in the analysis. Four hundred and seventy two titles were listed for comics published in more than one calendar year, and these were tabulated by years to give an index of long-lived comic books. The number of these current in each year 1935–1974 was calculated; it is here expressed in terms of percent of the mean (73.3) in print each year. Illustration 7.2 compares our index of total number of issues of science fiction magazines published in each year with the new index of numbers of comic book titles in print. Of course the comic book index begins with a rise, but the most noteworthy feature of the comic book curve is the great hump 1948–1955. This may be compared with the great boom in science fiction publishing in the same period and very strongly suggests that the great boom resulted from publishing market conditions and had nothing to do with an upsurge in popular interest in space.

The impression conveyed by the graph is supported by a simple mathematical analysis. The correlation between the two indexes for the

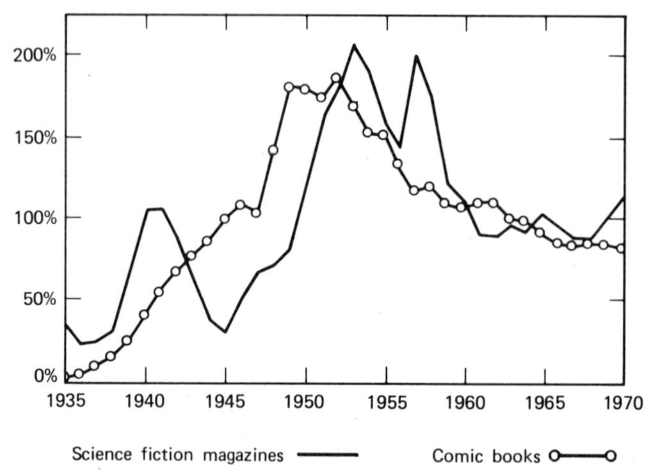

Illustration 7.2 Science fiction and comic books. Issues of science fiction magazines published each year and estimated number of long-lived comic books in print, expressed as percentages of their mean values for the entire period.

31-year period 1940–1970 is $r = +0.52$—rather high, considering that the comic book index is only an indirect measure of the number of issues published for each year. This positive correlation is not explained by any simple secular trend affecting both indexes because trend lines for both are almost exactly horizontal.

Fandom

Over the same period the network of fans and authors grew and developed into the well-organized subculture that exists today. There is a guild of SF authors, the Science Fiction Writers of America (SFWA) which holds an annual general meeting at which the Nebula awards for best writing of the year are conferred; operates a speakers' bureau arranging the authors' lecture appearances, publishes a newsletter, and embodies a social network of intimate friends and collaborating professionals. The authors, many of whom used to be fans, are typically strongly tied to the networks of current fans. Not only do authors appear consistently at fan gatherings, but a number of the authors have maintained their fannish activities throughout their professional careers. Of course many active fans labor to break into professional science fiction writing.

Large numbers of fans are members of clubs; the two Boston local clubs, The New England Science Fiction Association and The MIT Science Fiction Society each have about 200 members. Many fans publish amateur fan magazines, *fanzines,* and at any given time about 100 different titles will be in print.

In response to my 1973 questionnaire (Q-NESFA-73), NESFA members reported data indicating they each subscribed to a mean of 1.91 fanzines, other than the club newsletter, 68% getting the biweekly newsletter *Locus,* which I have found an invaluable source of information about events in the field. Respondents to my 1974 questionnaire (Q-FANZ) sent to fanzine editors revealed that these people read a mean of 14.0 fanzines a month and belonged to a mean of 2.9 SF clubs.

Not only do SF clubs hold frequent meetings, but for the last 35 years large-scale science fiction conventions have served as important meeting grounds for fans, authors, illustrators, and editors. The first World Science Fiction Convention was held in New York in 1939, and about 200 fans attended. A peak of 1,000 attendees was achieved in Chicago at the 1952 "worldcon" in the midst of the great boom. This figure was not surpassed until 1967 when 1500 fans gathered in New York.[24] Attendance figures for worldcons now typically surpass 2000, and around 4500 attended the 1974 convention in Washington, D.C. In my 1973 questionnaire sent to members of NESFA and my 1974 questionnaire distributed through a national sample of fan magazine editors, I asked respondents how many SF conventions they had attended in the past year and in their entire lives. NESFA members had attended an average of 2.23 cons in the past year and 12.85 in their lives. Fanzine editors and their associates reported 2.41 cons in the past year and 15.23 in their lives.

The standard deviation for the lifetime figure for Q-FANZ respondents is 19.41, indicating a great variation in convention attendance. To determine how many SF-cons a person could attend if he were really fanatical about them, I went through a large number of fanzines, culling information about cons held in 1972. There were at least 27 conventions held in the continental United States in that year, filling 69 different days and 21 of the 53 weekends with fan activities.

By various complex but not particularly interesting manipulations of convention attendance figures, I concluded that as of 1972, about 5400 Americans would have attended a world science fiction convention if it had been held within their home states. This can be taken as a rough estimate of the hard-core SF fans in the country at that time, although some rabid collectors do not participate in conventions. Even if we assume a more generous estimate of twice as many, or roughly 11,000, "trufen" (real SF fans) constituted only a tenth of the 110,000 circulation of *Analog*, the leading SF magazine. But how shall we define the true fan? Harry Warner has defined an SF fan as "the person who does something more about his interest in science fiction than the simple process of reading stories of this type."[25] Long-time fan and professional editor A. W. Lowndes says:

> What I will mean by "fan" is any person who, in addition to reading SF, does one or more of the following things: writes letters to editors, is in active correspondence with other fans on the subject of SF, is an active member of some fan group, contributes material to fan magazines, publishes fan magazines, attends or participates in conferences, conventions, etc.[26]

Fans attribute a number of qualities to themselves and to SF which stress the subcultural nature of the phenomenon and the relative deviance of fandom. In my questionnaire Q-NESFA-73 I asked respondents to agree or disagree with a selection of actual fan comments about fans, and despite the possible bias in the percentages, the results are interesting. In response to the comment that fans are "caught between the hip life and the straight, 66% agreed; 71% supported the statement that "fans are literate loners in an illiterate and group-oriented society. Having more imagination than others and holding views radically at variance with their own culture, they band together to find a little comfort." That fans are "shy, introverted types—tending toward the intellectual," evoked 73% agreement; and 71% said they "read science fiction because, in part, it is about people who are more than average, not people who are less."

In response to another open-ended question, fans contrasted themselves with average folk: "I think fans are more openly free in mental attitudes; they can accept a bit more easily that not everyone fits in the same mold nor belongs there." "Fans are more independent of mind. They question more." "I think fans are more imaginative and future-oriented, also more openminded." "Fans have a longer viewpoint than most of society." "We're more open to new ideas and experiences than most people...most of us are more aware, and more tolerant of new, strange ideas and beliefs." "Fans tend to be better educated," and of "above-average intelligence." "Most fen [fans] seem to be a least a little dissatisfied with the world as it is today, whatever day it is."

Fans make a clear distinction between the fannish world and the "mundane" world of average people, much like the hip/straight distinction made by hip members of the counterculture. A Boston fan reported a conversation between two mundane people, overheard at Noreascon, the 1971 world convention:

MUNDANE IN SHERATON
LOBBY TO
ANOTHER SAME:

> Who are all these ugly people? Why are they here?

HOTEL SECURITY GUARD:

> You shouldn't talk about them that way, they're really very nice people.[27]

One fan wrote: "Most fans *prefer* to be as strange as possible." Another argued: "Fandom is a peculiar sort of neurosis."[28] A science fiction fan is "a weird thing from Mars...that strange mythical being."[29] One of the 31 fans I interviewed at the 1971 Noreascon meeting, said that a fan

> ...as viewed by the outside world, is a strange, green funny-eyed being that sits around and watches these horrible movies late at night at drive-ins. And, normal people—in quotes—are not supposed to come close to them or do with very extreme caution. But, to the true fan, it's the most uh..it...being a science fiction fan opens up galaxies of thought and ideas and possibilities that are not available in the so-called normal world.

Continually, fans heaped scorn on mundane thinking and on the straight society of which many of them did not feel a part. One told me:

> I really don't think of myself as [an outcast]. I think of myself as an anomaly. But, outcast?...Since I would be being outcast from something I don't particularly like anyhow—and this is not really sour grapes—I [find myself] saying, "Okay, I'm anomalous; if you don't like it—tough!"

Another fan, who dressed and moved eccentrically and who appeared seized by notions of sinister plots all around him, both in his personal life and the world at large, said a fan

> ...is not totally satisfied with the life he is now living, and of an open mind to accept and utilize thought and thought construction of how it may be or could be. Nothing is accepted as absolute truth, and no form of life is accepted as absolutely desirable. We never believe in the best of all possible worlds. We're always looking for something better... Unfortunately, mainstream literature has a closed corporation, where everybody slaps everybody else on the back and criticizes everything they haven't written or don't like. Like, "If it's good, it can't be science fiction; if it's science fiction, therefore it can't be good, because X said so." As you'll notice in the Nielsen ratings [which are faked].

Leading editor John W. Campbell, Jr., also saw a great conspiracy behind the labeling of science fiction as bad literature and its supporters as deviant:

> The modern set-up in "literature" is that the term is restricted to things that meet the approval of the small, self-adulatory clique of Literateurs who have decided that they, and they alone, are fit to determine what is good and what is worthless. The number of those who constitute the Literateurs is remarkably small—but they are most remarkably effective in guiding the reactions of the Sheep of Suburbia. What they say is Bad, the sheep baah at faithfully—and they can do a remarkably effective job of lousing it up commercially. What they say is Good, the sheep ooh at and buy, even if it has no intrinsic merit.
>
> The Literateurs do not like any form of literature that incites the sheep to think for themselves—which is the avowed purpose of science fiction. They are, therefore, very ready to grasp any evidence that science fiction is bad.[30]

Focal Concerns of the Subculture

One way to approach the question of what fandom is all about is by examining how and what fans communicate to each other. Fan historian Harry Warner has said:

> One proof of the existence of a subculture is its consistent use of its own dialect or a plentiful supply of slang. Fandom's slang is used consistently

> enough to cause bafflement and resentment in many individuals who encounter a fanzine or attend a club meeting for the first time.[31]

Reasoning that a study of the special language of the subculture would reveal much about the concerns, orientation, and style of the group and its members, I conducted an analysis of science fiction language, "Fannish." It was not my purpose to produce a dictionary of all Fannish words ever used, but to collect a representative sample of the language, then inspect its focus and structure. Because much of the communication between SF fans takes place through the medium of mimeographed fan magazines, I tabulated all special words that appeared in a selected sample of fanzines, totalling 1621 printed pages.[39]

A very large number of words appeared only once, and these were rejected because I believed most of them were coinages, not standard elements of SF culture. Four common special words were rejected because they were journalistic slang, not peculiar to SF: *info, illo, repro,* and *typo.* My final glossary contains 53 words with etymologies.

Something of the cultural style of fandom can be seen in the highly mechanical methods by which most of the words were originally created. Only six of the 53 words represent special imagery or colorful usage, for example *mundane* and *android.* One word, *grok,* may be the contraction of *grow close.* The remaining 46 words were all created by the contraction of conventional words, by running words together, or by the construction of an acronym. *Con* is short for convention. *Sercon* is the contracted combination of *serious* and *constructive. Apa* is the acronym of *amateur press association.* Although a few of the contractions show a certain imaginative flavor, for example *egoboo* from *ego boost,* almost all the words were produced mechanically. A metaphor is hard to find in the entire list. The lexicons of other special languages often contain colorful words created by rather more poetic processes than employed by fandom. Consider for example the evocative words of the drug subculture: *snow, grass, roach, joint.*

Fannish words are primarily about fanzines, conventions, and fan society. There are very few words about space, science, or indeed, about SF literature. Talk using these words is focused inward on the introverted concerns of fandom. The words do not describe activities or values aimed toward the outside world or toward change. Interestingly, although science fiction authors habitually coin many words for use in their stories, very few of them achieve general currency in fan communication.

Another way of delineating fandom's concerns is to look at the subjects discussed when fans gather. A standard feature of large science fiction conventions is the informal seminar, organized around a specific topic and held typically in a hotel room. Sign-up sheets for such ad hoc seminars are posted publicly. At the 1972 Los Angeles world convention, L. A. Con, I copied down all the titles of the 48 seminars advertised and counted the number of names subscribed on each. There is usually an upper limit on the number of people who can attend, and individuals may have signed up for more than one; so this is not a perfect index of interest, but we certainly can learn something from a comparison.

I divided the 48 seminar topics into seven rough categories. Table 7.2 summarizes the data by category. Only three of these seminars could even approximately be described as pertaining to science and

TABLE 7.2 L. A. CON DISCUSSION GROUPS–SUMMARY OF TOPICS AND CATEGORIES

Category	*Largest Groups (attendance in parentheses)*	*Groups in Category*	*Number of Names*	*Percent Subscribed*
Science and technology	"How to Hijack the First Starship" (30)	3	48	6.2%
Literary, academic	"The Teaching and Criticizing of SF" (44) "SF and the University" (24)	6	110	14.3%
Author or book	"The Writing of Harlan Ellison" (52) "Katherine Kurtz Reading" (39)	12	259	33.6%
Film and television	"The Star Trek Phenomenon" (47)	6	136	17.6%
Fandom	"The Con Game" (24)	5	49	6.4%
Related genres and media	"Underground Comix" (34)	9	84	10.9%
Writing technique and Criticism	"Mapping Out a Novel" (22)	7	85	11.0%
Total		48	771	100.0%

technology: "The Reality Interface" (computer oriented), "The Predicting of Future Technologies," and "How to Hijack the First Starship." By far the most popular category concerned specific authors and books, and clearly the majority of the seminars were focused on topics relevant to science fiction literature and related genres. Sometimes an author would be the guest of a seminar about his own works. The presence of authors at conventions of this type orients fans toward them and their words. Because fans are deeply engaged in *fanac* (fan activity) at the convention, they feel little inclination to attend seminars merely talking about fanac. The second least popular category concerned fandom.

Another plausible index of fan orientation is provided by the costumes worn by fans in the masquerade contests held at major conventions. I photographed almost all the costumes shown at the 1971 and 1972 world conventions, a total of nearly 160 different costumes. I even observed a couple of important authors, Larry Niven and Lin Carter, dressed in unusual attire. Niven was pretending to be a magical mathematician; Carter was going as a character he was currently writing, the Wizard of the Red Flame. Many of the men went as ferocious monsters, warriors, or barbarians; a number of women dressed as their victims or as slave girls. The costumes did not seem to include many that could conceivably be called "scientific." To secure independent opinions, I presented 50 appropriately selected slides of costumes to 16

*TABLE 7.3 PHOTOGRAPHS OF 50 SCIENCE FICTION COSTUMES JUDGED BY 16 CODERS**

Category	*Number*	*Percent*	*STRANGE*	*LIKE*	*DIONYSIAN*
Scientific	32	4.0%	3.56	2.69	2.44
Erotic	34	4.3%	3.53	2.76	3.47
Military	38	4.8%	3.50	1.76	2.11
Religious	53	6.6%	3.51	2.49	3.04
Ethnic/Foreign	55	6.9%	2.82	3.31	3.25
Supernatural	89	11.2%	4.53	2.48	3.03
Historical	97	12.2%	3.13	2.87	2.85
Other	125	15.7%	2.75	2.45	3.10
Fantasy	275	34.5%	4.07	2.76	3.31
Total	798	100.2†	3.61	2.66	3.08

*(Mean scores: 5 = high; 1 = low)
†*Rounding error.*

Wellesley College students and asked them to assign each to one of nine categories and give it a score on three "dimensions": Everyday/Strange, Apollonian/Dionysian, and a preference scale. Table 7.3 gives the results.

The sample of slides contained 23 (45%) of male contestants, and 27 (54%) of female contestants. Photos of the females were judged more Dionysian than the male photos. The percentage of "erotic" costumes was underestimated, because I decided not to include any photos of topless or completely naked females in the sample shown to the 16 female coders to avoid offending them. Even with this underestimate of the "erotic" category, the "scientific" category is chosen least often.

Science Fiction Inventions

I have heard many times that SF authors often write stories containing brand new technical ideas and are therefore not merely oracles, but bona fide inventors. Writing in *New Scientist,* Donald Gould has even suggested, somewhat jokingly, that science fiction writing should be regulated by an "international Society for Social Responsibility in Fiction Writing," because otherwise much dangerous technology might be spawned by the uncontrolled imaginations of the writers.[33]

In questionnaire Q-NESFA-74 I presented members of the Boston area SF club with a series of statements and seven-point scales on which they were to indicate the extent of their agreement or disagreement. One was "Science fiction generates many new technical ideas which engineers later develop into practical inventions." Of 64 respondents, 56.3% agreed, 31.3% disagreed, and the rest had no opinion.

I also asked: "Do you happen to know of any specific scientific or engineering idea of value which first appeared in a science fiction story?" Of 65, 17 (26%) said they did not, while the 74% majority said they could think of such an idea. I then asked them to describe the idea and name the author and story from which it came. Many ideas were named without any documentation: submarines, rockets, computers, satellites, test-tube reproduction, genetic engineering, atomic bombs, nuclear power, and television.

Several ideas were described by one respondent each. The idea of using thermonuclear devices for urban renewal, suggested in a story by Tony Lewis, was given as an example of a valuable technical idea by one person who, surprisingly, described his political orientation as "moderate!" One said Jules Verne had invented the air lock in *20,000 Leagues Under the Sea,* and another mentioned Verne's general discussion of underwater conditions and technology. Other ideas mentioned and documented to some degree were the hydrogen bomb, the military tank, "belt-way" transportation, and space stations.

Two ideas were mentioned with great frequency; 19 named Robert Heinlein's invention of "waldo" remote manipulation devices, and 13 named Arthur C. Clarke's invention of the synchronous communication satellite. In the novelette *Waldo,* Heinlein described a physically feeble mad scientist who had devised mechanical arms with which he could perform feats of power and dexterity. It is often contended that the remote manipulators used today in atomic laboratories, and occasionally called waldos, are derived from this story. Heinlein himself has written that he got the idea from a 1918 copy of *Popular Mechanics* which contained an article about a feeble engineer who in fact had fashioned similar mechanical arms for himself.[34] Heinlein commented:

> ...much has been made of the "successful prophecies" of science fiction—the electric light, the telephone, the airplane, the submarine, the periscope, tanks, flamethrowers, A-bombs, television, the automobile, guided missiles, robot aircraft, totalitarian government, radar—the list is endless.
>
> The fact is that most so-called "successful prophecies" are made by writers who follow the current scientific reports and indulge in rather obvious extrapolation of already known fact.[35]

That Heinlein and another author, Cleve Cartmill, "invented" the atomic bomb in their stories has been suggested. In the 1941 work "Solution Unsatisfactory" and the 1940 story "Blowups Happen," Heinlein "predicted" atomic power plants and bombs. He says he got the basic ideas from technical journals and from conversations with scientists actually working in the field.[36]

In 1944 John W. Campbell's magazine *Astounding Science Fiction* published a story by Cartmill which described a future atom bomb in some detail. Several people in the U.S. Government familiar with or actually involved in the atom bomb project became alarmed and apparently believed Cartmill was betraying secret information. An agent was sent to interrogate the author. This incident itself became the subject of a later science fiction story which was produced as a television play; and in the traditions of fandom, the incident was inflated out of all proportion. Campbell for years used the Cartmill affair as proof of the significance of SF. Cartmill, of course, had merely based his story on already published reports on the feasibility of atomic weapons.[37]

The Arthur C. Clarke case is rather more complicated. In an essay subtitled "How I Lost a Billion Dollars in my Spare Time," Clarke has recounted the history of this invention, humorously lamenting the fact that he failed to obtain a patent which he could have used to extract great sums from Comsat and various world governments.[38] Although Clarke later became aware of communication satellite ideas put forth in 1942 by SF author George O. Smith and in 1923 by Hermann Oberth, his 1945 article "Extraterrestrial Relays" really does seem to have been the publication that both perfected and announced the idea. Clarke was not familiar with the earlier thought on the subject until after he had come to the invention himself.[39] As it happens, he presented the idea in an article in an electronics magazine, not in a story, acting in his capacity as leader of the British Interplanetary Society, not as an SF author when making the invention. Further, Clarke recognizes:

> ...If I'd not published in 1945, someone else would have done so by 1950 at the very latest. The time was ripe and the concept was inevitable; it was certainly bobbing around in the back of many ingenious minds.[40]

Jules Verne is often cited as the inventor of new technical ideas. Both the submarine[41] and the special breathing apparatus described in *20,000 Leagues Under the Sea*[42] were not original with him, but already existed under just the names he gave them. Verne, like many later SF authors, merely publicized technical achievements with which the general public was not yet familiar. All the authors added was a considerable but simple inflation of the inventions' capabilities. The verisimilitude of his books about a trip to the moon was assured by the aid of his brother-in-law, a professor of astronomy, who made the necessary calculations.[43]

After considering quite a number of SF "inventions," I have come to the conclusion that most of these ideas were derived from three kinds of sources: (1) earlier science fiction (2) current newsworthy science and (3) the occult. One recent novel incorporates an idea derived from all three kinds of source, so I shall discuss it as the most congenial example.

Larry Niven's award-winning 1970 novel *Ringworld* takes place on an artificial planet of unusual design. A band of fantastically strong metal has been forged around a star, with a radius similar to that of Earth's orbit around the sun. Ringworld spins at a velocity of over 1000 kilometers per second, and the resulting centrifugal force gives an artificial gravity in which *down* is *out,* away from the center. The inside surface of the ring is provided with soil, seas, mountains, forest, and all the other features of Earth's biosphere. While the inhabitants of Earth all stand with their feet toward the center of their planet, the inhabitants of Ringworld all stand with their feet pointing away from the center of theirs. The atmosphere on the inside of the ring is kept from spilling over the edge by a high wall. The surface of the inside of the band, the habitable area of Ringworld, is 3,000,000 times that of Earth.[44] The probable sources of the Ringworld concept are discussed below.

In 1818 Captain John Cleves Symmes of the U.S. Infantry presented his theory that Earth contained four undiscovered spheres, concentric with the familiar outer one, which could be reached through openings at the poles.[45] Throughout the nineteenth century, other writers supported Symmes' crank notions and gave further publicity to the idea. In 1913 Marshall B. Gardner presented a somewhat different hollow-earth theory. He held that Earth is a single hollow sphere with an inner sun that illuminates the habitable inner surface.[46]

Edgar Allen Poe wrote at least three stories influenced by Symmes, most particularly "The Narrative of Arthur Gordon Pym," published in 1837.[47] Jules Verne, who was greatly influenced by Poe and continued some of his stories, published his related novel *A Journey to the Center of the Earth* in 1864.[48] George H. Waltz, one of Verne's biographers, comments: "But one thing is certain, science to Verne was the essence, while to Poe it was like a fine powdery dust to blow into the eyes of the reader to make or to conceal a point."[49] I. O. Evans, another Verne biographer, reports:

> Though not the first to make fictional use of the hollow-earth theory, Verne was the first to place his narrative on a scientific basis. He gained much technical information from the French seismologist Charles Sainte-Claire Deville, and he had heard it suggested that the Italian volcanoes might be connected by underground fissures. Why should not these lead to an immense subterranean cavity which he could people with the extinct animals of the past?[50]

If Verne moved in the direction of science, Edgar Rice Burroughs turned back toward pseudoscience for his *Pellucidar* series of novels, published from 1913.[51] In Pellucidar dwell all the species of animal that ever existed on Earth, protected from extinction by the unchanging climate of the inner world. Although inspired by Verne, Burroughs has designed Pellucidar along the lines of Gardiner's hollow sphere.

In Ringworld, centrifugal force acts as an artificial gravity, pressing its inhabitants out against the ring; in Pellucidar there is no plausible gravity at all, and contrary to the testimony of the stories, the inhabitants would have floated uncomfortably about in the giant hollow. The earth above their heads would tug with sufficient gravitational attraction exactly to balance out that below their feet. Burroughs recognized the impossibility of his fictionally useful world, and chose its name to convey his awareness. *Pellucidar* comes from the adjective *pellucid,* meaning "easy to see through."

Burroughs made more money from his books, which included the popular Tarzan series, than any SF writer since, and his works are extremely familiar to fans. Some of the most active clubs and fanzines are dedicated to his works, and the Burroughs Bibliophiles is one of the few groups to hold mass, regular meetings at the major conventions. Larry Niven came up from the fan ranks, remains a superfan who puts many "in" references into his stories, and is married to the legendary lady fan, "Fuzzy Pink." That Ringworld is not in great measure derived from Pellucidar is inconceivable.

Although Niven declined to grant me an interview, I was able to overhear him discussing the derivation of Ringworld from the so-called "Dyson Sphere." In 1960, following a line of thought explored by Tsiolkovsky many years before, physicist Freeman J. Dyson suggested that in principle a hollow inhabited sphere could be constructed around a star so that all of its energy could be used. He asserted:

> Malthusian pressures will ultimately drive an intelligent species to adopt some such efficient exploitation of its available resources. One should expect that, within a few thousand years of its entering the stage of industrial development, any intelligent species should be found occupying an artificial biosphere which completely surrounds its parent star.[52]

In reply to various arguments about the feasibility and stability of the huge hollow sphere implied in his essay, Dyson agreed that a solid shell around a star was impossible and said he had meant to describe a "loose collection or swarm of objects"[53] But Dyson's idea entered science fiction as a sphere.

Niven met the objection that such a sphere, only 2 to 3 meters thick, could not maintain its form by postulating new materials for its construction and met the objection lodged against Pellucidar's lack of gravity by transforming the sphere into a ring and making it spin.

The idea of providing space ships with artificial gravity generated by rotation is not new. Tsiolkovsky discussed it in 1895.[54] The first real spaceship to experiment with artificial gravity was *Gemini 11,* launched in 1966. It achieved only 1/700 of the force of gravity at Earth's surface using an Agena vehicle at the end of a 30-meter cable as a counterweight for its rotation.[55] Ringworld is more directly derived from the toroidal space station design popularized by von Braun and other authors in 1952.[56] The original of this plan for a rotating, ring-shaped, 50- to 100-meter-diameter space station was described by "Hermann Noordung" in his popular book *Das Problem der Befahrung des Weltraums,* published in 1929.[57] Noordung is a shadowy figure; several sources say his real name was Potocnik and that he was a Captain in the Austrian army.[58] An article by this popularizer and inventor of space technology, "The Problems of Space Flying," appeared in the July 1929 issue of the science fiction magazine *Science Wonder Stories.*[59]

To claim that SF writers of the early days described dashing heroes who jaunted about the solar system in rocket ships is incorrect. Before real space pioneers had demonstrated that the rocket was the most feasible vehicle for leaving the earth, SF authors had their heroes travel by the most diverse means. For example:

1. Being shot from a huge gun. Jules Verne's astronauts use this dramatic method, as did H. G. Wells' Martians.
2. Anti-gravity devices and substances. H. G. Wells' *The First Men in the Moon* discover a material, cavorite, with negative weight that lifts their craft into space.
3. Electric devices. Hugo Gernsback's *Ralph 124C41+* travels to Mars in one.
4. Occult rays, and so on. The "eighth ray" of light propels ships from Mars to Phobos in Edgar Rice Burroughs' *Swords of Mars.*
5. Transmigration of souls. Burroughs' heroes in *A Princess of Mars* and *The Master Mind of Mars* die on Earth and have the good fortune to be reincarnated on the red planet.

Using *Day's Index* I tabulated the frequency of appearance of the word *rocket* in science fiction story titles from 1926 to 1950. Words in titles of

TABLE 7.4 FREQUENCY OF THE WORD ROCKET *IN SF STORY TITLES*

Years	*Number of Stories with* Rocket *in Title*	*Number of Such Stories per Magazine Issue*	*Ratio of Such Stories to Total Estimated Stories*
1926-1930	0	0	0
1931–1935	1	0.0054	0.0009
1936–1940	7	0.027	0.004
1941–1945	7	0.021	0.003
1946–1950	8	0.020	0.004

SF stories do in fact vary with time in meaningful ways that give a crude index of interest in the subject named by the word. For example, I tabulated *air* and *flight* and found that exactly as many titles (17) used these words in the aviation-conscious 1926–1930 as in the entire period 1931–1950! The data for *rocket* are given in Table 7.4.

The word *rocket* is uncommon before the mid-1930s, enters, then achieves and holds a constant level. Indeed, the concept of *rocket ships* seems to have been introduced directly into SF magazines by articles published in them by pioneers such as Max Valier, Hermann Noordung, and P. E. Cleator.[60]

Fan Support for Spaceflight

Although science fiction may not contribute valuable inventions to the Spaceflight Movement, it may serve to mobilize public support to vote and pay for space progress. On the face of it, this hypothesis looks like an obvious truth, but in fact there is some room for doubt. Some professional aerospace men fear that science fiction may work against actual space development in the long run by giving the general public such distorted images of what is possible that real successes will not be appreciated.[61] The stigma attached to early science fiction may have contaminated the Spaceflight Movement in the public's mind; they could not have seen rocket development as "crazy Buck Rogers and Flash Gordon stuff" had there been no such SF stories. We do not have the data to test the broad hypothesis that science fiction has served the Movement by effectively spreading pro-spaceflight values. We can, however, look at the more narrow question of the level of support given space development by contemporary SF fans.

At the world science fiction convention held in 1971, fan Guest of Honor Harry Warner used his brief speech before the awards banquet to urge active fan support for the space program. At one point he said: "If the comments I've been reading in fanzines are a good indicator, not more than perhaps 50 or 60 percent of all fans or professionals are solidly behind the continued exploration of space."

In the 1973 NESFA questionnaire I asked fans to comment on this statement. Twenty of the 71 respondents could neither agree nor disagree with Warner, and of the others, 45% (23) agreed and 55% (28) disagreed. The form of the question may have produced a positive response bias, and what fans would understand "solidly behind the continued exploration of space" to mean is not clear. Two important possible interpretations are: (1) "willing to say that the space program should be expanded or vigorously continued" (2) "frequently engaged in public activism or other action aimed at furthering the exploration of space." We shall look at these alternate definitions in turn.

In early-1969 the Gallup Poll asked the following question of its national sample of citizens: "The U. S. is now spending many billions of dollars on space research. Do you think we should increase these funds, keep them the same, or reduce these funds?"[62] I included a similar question in the questionnaire I sent to NESFA members in 1974 and another administered to an accidental sample of participants in the 1975 "Boskone" New England regional SF convention. As can be seen from Table 7.5, science fiction fans are much more likely than average citizens to say the space effort should be expanded.

In Q-NESFA-73 and Q-CFF I included another question asking what NASA's dollar appropriation should be for 1974. Of those who re-

TABLE 7.5 RESPONDENTS' OPINIONS ON WHAT SHOULD BE DONE WITH SPACE APPROPRIATIONS

	Gallup 1969 (N = ca. 1500)	*NESFA 1974 (N = 81)*	*BOSKONE 1975 (N = 79)*
Increase	14%	84%	80%
Keep same	41%	10%	6%
Reduce	40%	6%	9%
No opinion	5%	0%	5%

sponded, 93% of the NESFA members and 62% of the CFF sample called for an increase. Although the then space-boosting Committee for the Future gave greater verbal support to NASA than the average citizen, the SF fans were even more solidly behind NASA. Among the CFF respondents, those who called for an increase rather than a decrease in NASA appropriations were more likely to have read more than one SF novel during the previous year. The difference was significant at the 0.01 level.

In 1972 sociologist Irene Taviss reported on a survey of public attitudes toward technology and mentioned a test of attitudes toward government programs:

> In order to determine the relative priorities among technological and social problems, we asked our sample to rank order the following programs: space program, pollution prevention, national defense, urban housing, mental health, welfare and poverty programs, and crime prevention. The highest priority went to welfare; the lowest, to the space program. National defense was ranked next to last, and pollution prevention in the middle, after crime prevention and mental health. The social programs were clearly ranked ahead of the technical programs.[63]

I included a question based on Taviss' in the questionnaire sent to fanzine editors in 1974 and in Q-NESFA-74. Taviss had drawn her sample from average citizens in the Boston area; in my two questionnaires I had a national sample of fans and a Boston-area sample. In terms of average rank, both fan samples gave pollution prevention first place, the space program second, and crime prevention third. Pollution prevention is a science-related problem of high consensus, so obviously fans ranked it highly. Another question asked the national fan sample to describe themselves as political radicals, liberals, moderates, or conservatives. Fan radicals and liberals gave the space program second rank, on the average; the moderates and conservatives ranked space first. Irene Taviss' citizen sample placed space last, so my data prove that science fiction fans are strongly in support of the space program.

Given that fans support the space program with their words, when asked about it, do they act upon their favorable attitudes and really perform as activists within the Spaceflight Movement? In a word, no.

One of the very few times in recent years fans seem to have attempted public action in support of any cause was when numbers of them across the country rallied in support of the *Star Trek* television series, which was in danger of being taken off the air after 1967. Fifteen people from NESFA engaged in a demonstration in front of the offices of WBZ-TV in Boston;[64] Cal Tech students picketed the NBC office in Burbank; and the network received an estimated 114,667 letters demanding a continuation of *Star Trek,* not all, of course, from true SF fans.[65] The campaign did not arise spontaneously from fandom, but was engineered by Gene Roddenberry, the producer of the show, and Bjo Trimble, an SF fan who had become an informal member of his team.[66] The fanzine *Inside Star Trek* was a professionally produced piece of Roddenberry publicity, and the subscriber received a form letter priming the reader for "Twisting the Peacock's Tail"—that is, for putting pressure on NBC.

While doing field research in NESFA I noted an unsuccessful attempt on the part of some fans to collect enough money to pay IBM to produce a typewriter printing ball master bearing the Runish fantasy alphabet invented by J. R. R. Tolkien. Many fans owned suitable IBM typewriters which they used to type their fanzines. Some thought it would be fun if they could buy Runish type balls, which might cost $50 each after an initial cost of $1500 for the master.

In 1971 a NESFA member attempted without success to stir up fan activism behind an effort to convince the U.S. Post Office to issue a commemorative stamp in honor of the Late John W. Campbell, Jr., long the influential editor of *Analog.*

Shortly before his death, Campbell launched his last special project, which seems to have died with him. In association with author Gordon Dickson and illustrator Kelly Freas, Campbell wanted to whip up public support for the fading American space program.[67] Freas did advertise a series of posters propagandizing for space through the fanzines, but no hint of his project seems to have reached the general public, who presumably were its object.

For two years after Harry Warner's 1971 call for a campaign of fannish activism in support of spaceflight, I read many fanzines, including the comprehensive newsletter *Locus,* and detected absolutely no consequences of his plea.

Other attempts to mobilize fans on behalf of concrete future-oriented projects have not proven successful. At the 1974 World Science Fiction Convention, the World Future Society maintained a literature table. It hoped to recruit some fans to its ranks, or at least sell them subscriptions to *The Futurist.* Part of the time this table was manned by

Ed Cornish, founder of the WFS, at other times it was left untended. He told me fans showed very little interest in the WFS or the literature he was giving out. I suggested that fandom might be uninterested in taking action to bring SF ideas to realization. He agreed and said the WFS had empirical evidence in support of this. The society had placed an advertisement in one of the major science fiction magazines; the number of responses was precisely *zero*.

If we accept the description of fandom and SF as a subculture, intellectual movement, or perhaps countercultural network, it must be further described as *extremely retreatist.* There is no evidence of any pro-space group emerging out of science fiction over the past 25 years.

The retreatism of fandom has deep historical roots. On March 10, 1935 SF fan William S. Sykora launched four small rockets he had built, undoubtedly solid-fuel rockets, as one of the activities of the small New York fan group, the International Cosmos Science Club. Throughout the 1930s Sykora was a major figure in fan politics, attempting to transform fandom into a propagandizer for science. To make fandom specifically into a space-boosting movement was not his purpose, but like editor Hugo Gernsback before him, he saw SF as a means for bringing young people to choose scientific careers and support a general increase of the influence of science in society. He found few allies within fandom and failed in his attempt to energize the growing subculture with proscience activism.[68]

A more powerful movement within fandom was the politically leftist, even avowedly communist, *Michelism* and the group of fans and authors who came to be known as the Futurians. John B. Michel, an older fan (already out of his teens), was a member of the Young Communist League. In 1937 he introduced his ideas for social revolution in deceptively technocratic language, calling for fandom to

> [P]lace itself on record as opposing all forces leading to barbarism, the advancement of pseudo-sciences and militaristic ideologies, and shall further resolve that science fiction should by nature stand for all forces working for a more unified world, a more Utopian existence, the application of science to human happiness, and a saner outlook on life.[69]

Although a number of fans supported Michelism, and even more associated themselves with the related Futurian club, the majority did not. Because the only common denominator uniting all fans was an interest in *SF as fiction,* an implicit collective decision was made to reject

Michelism's call for activism as a threat to fan unity. The Futurians were barred by force from the First World Science Fiction Convention in 1939, and similar groups have not challenged fandom's retreatism in an effective way since.[70]

SF: Breeding Ground of Deviant Movements

Although science fiction does not foster activism on behalf of any important conventional cause, it is an extremely powerful producer of occult groups, pseudoscientific sects, and other deviant movements. One interesting tiny movement with cosmic purposes recently spawned by science fiction is the Foundation, inspired by the trilogy of novels of the same name by Isaac Asimov. One fan I interviewed at the 1971 World Convention, praised

> Asimov's *Foundation* trilogy, which is probably...one of the pillars of science fiction, at least to me, shows how History and Psychoanalysis and things of this nature can be used to predict the future, possibly predict the future outcomes of civilizations.

Thousands of years from now, in the ruins of a Galactic civilization fallen into Dark Ages, a prepotent corps of social scientists called Foundation struggles to apply the discoveries of the sociological genius Hari Seldon, in the hope that civilization can be restored in 1000 years rather than the predicted 10,000. Critic Damon Knight has complained that Asimov's future society is merely a warmed-over projection of the Roman Empire, and at no point are Seldon's discoveries specified.[71]

Inspired by these books and in anticipation of the imminent fall of our own civilization, Gary Hudson, a college student, decided to assemble his own Foundation with the hope of creating a network of experts, enthusiasts, and institutes dedicated to protecting our cultural and scientific heritage and working through advanced technology for a positive future. Hudson never attempted to state the aims of this Foundation in any concise manner. He told me that the best description of his attitude toward the direction Foundation should go was expressed in a fictitious "Seldon's first null hypothesis," provided by an anonymous correspondent: "A system undergoing evolution cannot derive a formalism to express that evolution."

Foundation must be described as a highly fanciful enterprise, undertaken by young people of little means and low status, decked out in the trappings of a cosmically important movement, replete with rocket-and-sunburst emblem, charts of imaginary interlocking organizations, and fancy personal titles, like "First Speaker." The key mythology of the group is millennarianism in modern dress: The world will be shaken, the current social order toppled, and Foundation will emerge as the most potent guiding force for all civilization. Among the suggested projects for the group was the construction of a "sietch," a "place of refuge in time of danger," as described in Frank Herbert's SF novel *Dune*. Included in Hudson's plans were a Foundation Institute that would perform research intended to solve quickly the problem of aging and a library that would preserve Culture against Armageddon.

The major project completed by Foundation was the publication of two issues of *The Foundation Journal,* a handsome large-sized magazine patterned after *The Whole Earth Catalog* and containing a number of futurological articles. Hudson never solved the problem of how to distribute the journal, and four months after the publication of the second issue, 6000 of the 10,000 copies remained. According to its editor, Susan Chamberlin, this burdened them with a debt of $17,000. Most of the costs had been met with borrowed money.

For a time Hudson hoped to recoup his losses and provide funds for his many other projects by making a professional motion picture based on Herbert's *Dune*. The Foundation did not have a formal membership structure. The active core of the group was Hudson and Chamberlin, assisted by about a dozen friends and encouraged by a total of 45 interested persons, who did not include many with scientific training.

Hudson spent the summer of 1972 attempting to write a screenplay for *Dune* and collect support, which included an abortive attempt to get pop singer Judy Collins to donate the music in exchange for letting her younger brother play the hero, presumably also for free. Chamberlin prepared to write a book about the creation of the movie. Their plans for assembling a cast, production crew, and cash were primitive and fanciful, and the project came to a vague end shortly thereafter. In January 1974 Hudson, styling himself "Chancellor of the Foundation Institute," was able to arrange a display at the Science Museum of Minnesota titled "An Integrated Space Transportation System." At the present time, it is not clear that Foundation exists outside its Chancellor's mind, and it has not become a significant social force boosting the space program.

Many classic science fiction stories deal with the concept of superhuman genetic mutants, a kind of master race hidden among the mass of humanity and slowly gaining power. One of the most influential SF novels, *Slan* by A. E. van Vogt, stimulated a good deal of discussion within fandom and encouraged a number of fans to wonder if they themselves might be unrecognized slan supermen.

The typical classic science fiction hero does not achieve his greatness; it is bestowed upon him. Special mental or characterological powers, even magical talents like telepathy, are important attributes of the standard SF hero, and his adventures constitute not areas for accomplishment, but tests of his superior qualities. While fans may be concerned with social status, my ethnographic research indicates that not many of them are enthusiastically engaged in the effective pursuit of status through achievement. The fact that only retreatist or magical social movements have been spawned by SF in recent decades squares with the idea that fans seek vicarious gratification through the stories rather than actual success through achievement. The slan fantasy can be seen as the desire for status acquired magically, without the necessity of active striving. As one fan put it: "We all like to believe that we are the long lost prince or a fairy changling. 'Fans are Slans' is merely another change on this old bell." SF author Brian Aldiss has suggested that the real meaning of *spaceflight* for fans was as an expression of personal magical fantasies, rather than as a practical project to be undertaken in the real world:

> For several years before the first space rockets lumbered up to lodge a human being in orbit, the sf magazines issued propaganda on the subject; when success came, one might almost have been forgiven for believing that "thinking made it so." Space flight had been the great dream, the great article of faith; suddenly it had become hardware, and was involved in the politics of everyday life....
>
> Space travel was a dream, the precious dream of sf fans. It was part of the power fantasy of the sf magazines. When space travel became reality, the dream was taken away from them. No wonder that the sales of magazines dropped dramatically after that! Commentators have always had difficulty explaining the fact, but the reason is simple—withdrawal symptoms were going on.[72]

Aldiss may not have explained the drop we noted in SF magazine publishing, because the drop happened before the first major space successes. However, as Aldiss suggests, *spaceflight* has in some powerful sense become alienated from science fiction. The arduous, obsessive, hard-hardware nature of real space exploration does not harmonize with the flamboyant, seven-light-years-in-a-stride image of space travel in the fiction.

In its short history the Committee for the Future dropped away from *spaceflight* toward seemingly direct indulgence in social and mystical activity. In science fiction as well there is frequently a short-circuiting of the chain of commitments that bridges the gap between the magical orientation of the primitive inner self to those activities that are objectively effective, according to some reality principle. The hero of Alfred Bester's *The Stars My Destination* develops a magical ability to fly to the stars without benefit of spaceship; the hero of Heinlein's *Stranger in a Strange Land* can "discorporate" his enemies, wish them out of existence. The slan is superior because he is born that way, indistinguishable from average folk except for the presence of metallic threads in his hair and the ability to read minds.

In 1945 fan Claude Degler issued a manifesto calling for the formation of a Cosmic Circle of superior science fiction people, essentially slans, who would create a movement and develop a program for the advancement of humanity. Degler seems to have been an obscure, gawky, somewhat rustic figure with none of the attributes that might have made him a serious candidate for the status of slan. SF historian Harry Warner says that Degler did fandom a service by enacting an unintended debunking parody of the "Fans are Slans" idea. "The absurdity of the proposition that others had hinted at so delicately became quite obvious when Degler spoke it out so boldly."[73] He did succeed in getting temporary support from superfan Forest J. Ackerman, who was then calling himself "4SJ" in slannish fashion, but in short order The Cosmic Circle collapsed and Degler was driven from the field. In Q-NESFA-73 I asked respondents to comment on the following anonymous quote, which was actually a part of Degler's manifesto:

> Man is evolving toward a higher form of life. A new figure is climbing upon the stage. Homo Cosmen, the Cosmic men, will appear. We believe that we are mutations of that species. We are convinced that there are a considerable number of people like ourselves on this planet, if only we could locate and get in touch with them. Someday we will find most of them, and then we will do great things together.[74]

An overwhelming 96% of those who commented rejected the proposition. Eight recognized the statement as a "Deglerism" or "Fans are Slans" propaganda. A number described it as "bullshit" or used a similar epithet; some found it ridiculous and even wondered if it were in fact a joke. Still, several qualified their negative reaction with some comment praising fans: "I think fans would have a higher proportion of high-IQ types"; "Many of us are smarter or more creative than the clunk on the street"; "Fans are less provincial, with a wider world viewpoint than most, and eventually society will be shaped along those lines...." Two commented that the author of the Degler statement must have been reading too much science fiction, and several implied that their own experience within fandom had disabused them of any suspicions they might have had that fans were slans: "I would have believed this *before* I entered fandom."

With more sociable than revolutionary intent, a fan group in Battle Creek, Michigan, decided in 1943 to create a Slan Center, an entire city block full of fans living together. The actual result was a single eight-room house full of fans, called Slan Shack, a rather more modest project which lasted only two years. One of Slan Center's proponents, Francis T. Laney, later expressed himself on the doctrine of the superiority of fans in a statement which I also presented for comment in Q-NESFA-73:

> If we fans had the necessary ability to be New Order leaders, we would be demonstrating that superiority in actual research and inventions, rather than spending our time reading escape fiction, publishing fanzines, and writing whacky letters. There are mighty few fans with the real ability to be any more than mere ciphers in our civilization. If fandom could take over and lead the way to the New Age, just what kind of an asinine, screwball world would it give us?[75]

The comments on Laney's statement were mixed. The general feeling of the replies suggested that many took Laney's strong language to be insulting, and some asserted that fans could hardly run the world any worse than the present leaders, even though they did not have the supernatural abilities of slans.

The most important movement that has emerged from science fiction in recent years is Scientology, a new and popularly successful religion created by SF writer L. Ron Hubbard. In the May 1950 issue of *Astounding Science Fiction,* edited by his close friend John W.

Campbell, Jr., Hubbard introduced a new psychotherapy technique called Dianetics. The first surge of interest in Dianetics came from science fiction fans, but thousands of ordinary people all across the country soon became involved in this novel therapy process.[76] Hubbard called Dianetics "a milestone for Man comparable to his discovery of fire and superior to his inventions of the wheel and arch."[77]

Standard medical authorities and psychiatrists branded Dianetics deviant. Some legal action was taken against Hubbard, but when he expanded his system into the religion called Scientology, the unusual beliefs and promises of the movement became protected and sacred. I performed extensive ethnographic research in the Scientology movement in 1970 and plan to write about it elsewhere, but one impression deserves to be reported. Scientologists take seriously all the most exciting and deviant ideas of science fiction. They believe in reincarnation, intergalactic civilizations, telepathy, force fields, immortality, levitation, and the possibility of training the mind until it can produce lethal beams of pure energy. To put it in the language of SF, many Scientologists believe they can actually become slans, through the Scientology "processing" techniques invented by L. Ron Hubbard.

Among the most successful fads within science fiction was the bizarre Shaver Mystery affair. Raymond A. Palmer, editor of *Amazing Stories* in the 1940s, published and supported a series of stories dressed up as factual articles by Richard S. Shaver. Beginning in 1945, these tales were presented as true reports of Shaver's adventures in caverns deep under the earth where a race of scientifically advanced gnomes, called *deros,* was preparing to destroy our civilization. Palmer claimed that the Shaver stories boosted *Amazing's* circulation from 27,000 to 185,000, but after a storm of fan protest against the hoax and under obscure circumstances, Palmer lost his editorial job.[78] He edited his own SF magazines for a while, turned to flying saucers, and later published several occult and marginal spaceflight books and the aerospace news monthly, *Space World.* In this case, as in the others I describe, an attempt to transform SF-occult speculations into reality was greeted by strong opposition within fandom.

In a review of Jack Williamson's popular SF novel *The Legion of Time,* Brian Aldiss wrote:

> Science fiction so often turns out to be a fairy tale—never more so than in this instance!...Like Doc Smith's [phenomenally popular Lensmen] saga,

this one also works on magic. Most traditional sf does so. The magical spells are given such names as "mentally released atomic power"; the hyper-drives light the way to Babylon.[79]

The Role of Science Fiction in Society

The best-known fan definition of science fiction, written by Sam Moskowitz, stresses the magical function of science within the fiction:

> Science fiction is a branch of fantasy identifiable by the fact that it eases the "willing suspension of disbelief" on the part of its readers by utilizing an atmosphere of scientific credibility for its imaginative speculations in physical science, space, time, social science, and philosophy.[80]

What have we found? Although science fiction played a vital role in disseminating ideas and values in the earliest years of the Spaceflight Movement, it drew apart from actual space developments in the 1930s. It does not produce many ideas useful in real spaceflight. We cannot be sure if it has an active role to play in producing a positive attitude toward NASA in the general public, and testing such a hypothesis would be difficult. The culture of science fiction is oriented toward magic as much or more than it is toward science. Although fans say they support the space program, their words are not transformed into action. SF does obviously serve to entertain large numbers of ordinary people, but if we are interested in social or technological change of any kind, a contended population is precisely what we do not need! Is there any other social or cultural function that our evidence suggests is fulfilled by science fiction?

Science fiction appears to function as what I call a *cultural redoubt.* I originally conceived this term when writing on the subject of *deviant knowledge.* I had noted that "false knowledge," such as pseudoscience, could often escape disconfirmation within the reality of the larger society by retreating into cultural enclaves maintained by marginal groups and institutions. A variety of social phenomena seem to protect deviant cultural elements from complete destruction in much the way that game preserves and remote geographical areas protect endangered species of animals.[81]

Science fiction contains and accepts many concepts strenuously rejected by standard science fact. Parapsychologists claim to be qualified scientists investigating real phenomena of telepathy, telekinesis, clairvoyance, and the like; standard psychologists disbelieve and debunk the very existence of the phenomena the parapsychologists study.[82] Science fiction accepts parapsychology as a literary device and actually seems more knowledgeable about and more favorably inclined toward this deviant pseudoscience than toward conventional psychology. In his history of *Astounding Science Fiction (Analog),* Alva Rogers asserts that "...sociologically oriented type science fiction...dominated the forties,"[83] and a wide range of SF stories are frequently described as sociological. What is meant is that they deal with social events and interactions, not that any hint of scientific sociology has found its way into the fiction. Today some authors do have some background in the social sciences, but for most of the history of SF *sociology* and *psychology* meant the *occult* and *parapsychology.*

Culture preserved in a redoubt may later emerge into the larger society. Spaceflight itself might be the best example of an idea protected and developed to some extent within science fiction and later brought to practical realization and full acceptance in the larger society. SF may contain other ideas that might not survive outside the redoubt at the present time, but may emerge strong and compelling at some future time when the conventional consciousness has changed.

Chapter 8 / THE FUTURE OF THE MOVEMENT

The next two decades will be a period of consolidation for the Spaceflight Movement. A number of types of near-earth or synchronous orbit satellites have already been developed and should achieve perfection in serving their various functions: communications, navigation, weather observation, earth resources research, environmental monitoring, and military reconnaissance. Highly specialized industrial processes may be carried out on a limited scale in the zero-gravity hard-vacuum conditions of space. There is serious talk of setting up electric power stations in orbit. NASA and Russian planners have scheduled a number of space probe launches over the next 20 years, although complete automated exploration of the solar system at current rates might take a century.

Many people imagine somehow that the space program is over, while really it is continuing at a fairly vigorous pace. It is the Spaceflight Revolution that is over, and the more normal period that has followed it appears one of inactivity only by contrast. The important question remains: Will there be a second wave of intense and dynamic progress? Will there be a second stage to the Spaceflight Revolution? William Irwin Thompson has painted a picture of the immediate future that would discourage any enthusiast for technology:

> The technology of our industrial civilization has reached a peak in putting a man on the moon, but, as the ancients knew, the peak is also the moment of descent. Before we ascend the next peak to Mars, there is a very dark valley waiting beneath us, and, poetically enough, its darkness is made up of just those things our civilization did in order to succeed. In straining our industrial technology to the limit, we have, in fact, reached the limit of that very technology.[1]

A number of authors have suggested that modern technology has reached either a natural scientific limit or the more social limit of having achieved all it is worthwhile to achieve.[2] Any argument over this question must be emotional and ideological. We do not have the facts to conduct a discussion on a more reasonable plane. In this chapter I consider a few of the technical and social opportunities for further progress in space—a more manageable but still rather speculative topic. There are several books on the market which present "gee-whiz" pictures of a future interplanetary civilization; where the authors run out of facts they substitute optimism and even discard scientific realities when they are uncomfortable.[3] Pessimistic books tend not to focus on the future in space, but on the future of technological civilization as a whole. Certainly, if general economic and scientific progress is at an end, space progress must also be. Here I make the assumption that terrestrial civilization will continue to advance in knowledge, power, and wealth, not because I can argue persuasively that it will succeed in doing so, but because the assumption is necessary before thoughts of extraterrestrial progress are at all plausible.

My overall contention is that the next 20 to 50 years will be marked by a gradual upward coasting of space technology capabilities—a period of normal technological change. Somewhere soon after the turn of the century there is the real possibility of a Second Spaceflight Revolution, or at least very rapid progress of the normal type. In the concluding pages of this book I consider the major areas in which explosive progress is possible and some of the social conditions that might facilitate it.

Discoveries and Inventions

The Spaceflight Movement could be significantly invigorated by new discoveries in space that would excite renewed interest in exploration and by new technical innovations that would reduce the cost of spaceflight.

The discovery that might give an explosive boost to NASA's funding would, of course, be the finding of an artifact left in our solar system by intelligent beings from outside. Undoubtedly men could not contain their excitement or their fear until the rest of our stellar neighborhood had been completely explored and brought under human domination. Occult groups and pseudoscientists have made a number of claims that such an object has already been located, but their evidence is worthless.[4] I think we have to make the conservative assumption that we will not be fortunate enough to find proof of extraterrestrial intelligence in this manner.

As I write, two *Viking* unmanned spacecraft are on their way to Mars. Part of their mission will be to conduct experiments on the Martian surface designed to detect the presence of life. A negative result would not be conclusive, but a positive result would be sensational. There is a nasty joke to the effect that the American space program sent a secret rocket ahead of the *Vikings.* Its mission was to put life on Mars so they could find it. By the time you read this book, the results should be in. Craig Covault has noted:

> NASA planetary officials believe that, if the Viking spacecraft finds life on Mars, substantial funding will be made available to follow the discovery with additional research as soon as possible.[5]

The discovery of purely scientific marvels in space is not sufficient to stimulate great investment. The findings have to be translated into relatively popular terms. Consequently, the most powerful impressions would be made by discoveries concerning extraterrestrial life. The front page of the *New York Times* for January 7, 1973 announced: "Scientists Feel Life is Possible on Saturn Moon."[6] A National Science Foundation handout proclaimed: "Astronomers Report Belief Atmosphere of Saturn Moon Similar to that of Earth,"[7] and *Aviation Week* informed its readers: "Saturn Moon Might Support Life."[8] The moon in question is Titan, largest of Saturn's satellites, which has long been known to have an atmosphere. Recent astronomical research suggested that Titan had more of an atmosphere than expected and might be considerably warmer than had been assumed. Different interpretations can be placed on the data currently available, and at best it could be said that the density and composition of Titan's atmosphere may indicate quantities of prebiological complex molecules could be found on the surface.[9]

This is neither the time nor place to come to a conclusion about Titan. Rather, we should note that Titan is being aggressively advertised as a place worth exploring. Further interesting discoveries, perhaps made by the *Pioneer 11* probe currently on its way to Saturn, or by one of the probes NASA plans to launch later on, could stimulate enough interest to get a Project Titan under way.

The technical innovations that would stimulate space development most markedly are still primarily in the field of propulsion. Later on, if sending payloads or spacecraft to the other planets becomes inexpensive, other kinds of innovation might become particularly important, such as improved life-support systems or efficient means for exploiting extraterrestrial resources. Future space technology probably will include two radically different kinds of vehicles: shuttle craft and deep-space craft.

We have already described the coming *Space Shuttle* designed to deliver men and equipment to Earth orbit on a regular basis and at lower cost than has been possible with nonreusable launch vehicles. Maybe the Shuttle will fail—either fail to meet its design requirements or lose federal support through a political reversal. But NASA's success record to date encourages me to believe that the shuttle will meet its objectives and become the standard orbital transport system for the last two decades of this century. Around the year 2000 a second-generation shuttle will probably be produced. Because of restrictions in the funding available to develop the first shuttle, it will not achieve the best economies possible today, so a second-generation shuttle would reduce costs still further, even if it embodied few technical innovations.[10]

New propulsion systems may be developed for later-generation shuttles. Environmentalists and the general public would raise a terrible protest if NASA ever tried to produce an atomic-powered shuttle. Nuclear fission rocket engines have been tested, and there is little hope at the moment that nuclear fusion rockets could soon be developed with high enough thrust for Earth launch. Consequently, an atomic shuttle means a fission Shuttle. I believe that nuclear fission technology is highly undesirable. At best it can be seen as a transitional technology that will be phased out in 50 years or so when fusion technology takes over. Currently there is a great debate over the safety of nuclear power. Even if some of the dangers of radioactive pollution have been exaggerated, any increase in terrestrial radiation levels is unacceptable.[11] It is certainly difficult to believe that NASA could sell the country on allowing it to build a nuclear fission shuttle.

Ever since the 1930s authors have speculated that chemical propulsion systems based on monatomic hydrogen might be possible, and some consideration has been given recently to metallic hydrogen as a propellant. Either would offer a much higher specific impulse than any existing fuel and in theory would permit the construction of neat single-stage shuttles with small fuel tanks, requiring no boosters. There are three related problems with these exotic substances: (1) No one has been able to make any. (2) If some were made, there is no known way to store it. (3) We do not yet have means for controlling the manner in which these two materials would revert to conventional hydrogen gas—that is, no way to keep them from exploding in the shuttle's fuel tanks instead of the engine. Various rumors hint that a breakthrough in one or the other may be just around the corner.

Throughout the early years of the Spaceflight Revolution, the proper means of assessing the efficiency of a launch system was with measures like specific impulse or mass ratio. The appropriate measure for space shuttle systems is not either of these, but that familiar indicator: dollars. Putting payloads into orbit is no longer very difficult; the problem is to do it cheaply. Therefore, exotic fuels may fail to replace conventional ones if their added energy is particularly expensive.

The main engines of the space shuttle burn hydrogen and oxygen, and there is no reason that they could not be the standard propellants indefinitely. Because they combine to form water when they burn, it is hard to imagine propellants less offensive to the environmentalists. Some writers have suggested that after the remaining petroleum has been used up, a "hydrogen economy" will dominate energy distribution.[12] Initial designs for airliners powered by liquid hydrogen have already been made.[13] At the present time most hydrogen comes from fossil fuels, but in the future electrolysis of water may become the major means of hydrogen production, also producing oxygen in the process.[14] If we look ahead to the middle of the twenty-first century, we could imagine a system in which hydrogen fusion reactors provide electric power to electrolysis plants which separate the hydrogen and oxygen in water and liquify the two gases for use as rocket propellants. Then the power system of a Space Shuttle would be analogous to an electric storage battery. Energy would be stored in the separation of the hydrogen and oxygen, to be liberated again in the rocket engine when they recombined. Both monatomic hydrogen and metallic hydrogen systems would also be energy storage systems. When gasoline is burned we are actually extracting energy directly from Nature; in these systems we extract the energy back in the fusion power plants, and the production of rocket propellants is merely a way of storing the energy in a useful form in the shuttle. The problems of production, storage, and handling have been largely solved for liquid hydrogen, and even if means were found for producing the other two forms of hydrogen, they would probably be more expensive. If my reasoning is correct, we cannot expect a major breakthrough in orbital launch systems but can look forward to a gradual reduction in costs as new-generation shuttles are built.

The situation is quite different for deep-space vehicles, rockets designed to be fired from Earth orbit to boost payloads on to higher orbits and the planets. First, several very good propulsion systems operate at great efficiency but produce very little thrust. Ion rockets, for example, can have specific impulses twice those of the best chemical propellants, but produce so little thrust that they could not lift themselves off the ground. Their tiny thrust can be kept up for very long periods—months or even years—so if they are fired in orbit where low thrust is no disadvantage, they are superior over chemical engines.

The best foreseeable deep-space propulsion system would be based on hydrogen fusion. Sebastian von Hoerner has estimated that a velocity of 20% of the speed of light might be achieved by intersteller probes propelled by the fusion of hydrogen into helium.[15] A more modest use for such a system would be transporting people and goods to Mars. The space shuttle would ferry everything into Earth orbit, and a fusion-powered deep-space tug could transfer the cargo to Mars orbit. Depending on what was most economical, the tug might even make several round trips before engine maintenance or even refueling.

There is no point in designing such fusion tugs today. We may be 50 years from the time in which one can be built. The reasonable thing for spaceflight engineers to do is wait until fusion reactors (of whatever type) have been perfected by atomic scientists and engineers, then begin to adapt them for space propulsion. Progress in fusion technology over the past 20 years has been painful and slower than many expected. We can hope that the first fusion power station will go into service around the turn of the century, with really widespread and efficient use of fusion power a generation after that. So somewhere between the year 2000 and 2030 we can expect the first successful fusion spacecraft. If economical fusion reactors are never successfully built, all bets are off, not only for spaceflight but for any kind of technological progress.

The fusion tug would be the essential link in an interplanetary transport system. If it comes into service, the economically limiting factor will not be transport between the planets. Rather, there would be two limiting factors: (1) the cost of shuttle service to Earth orbit and (2) the cost of scientific, life-support, and resource exploitation activities at the other end of the journey. Thus, the fusion tug would not in itself cause a golden age of interplanetary expansion. The other technologies would also have to advance, and there would have to be compelling social reasons for expanding the space program.

The Space Patrol

Is there any way the Spaceflight Movement could engineer a second "military detour?" It is both unpleasant and difficult to conceive of new military projects that could be used to further the development of spaceflight. However, because this is the route by which most of the progress has been achieved, I felt it was necessary to devote a good deal of thought to this issue. Most ideas along these lines turn out to be valueless. For example, a useful military base on the moon, complete with soldiers and civilian technicians, used to be remotely conceivable. Now that Earth-based ICBM's have been developed, the moon base could only provide some kind of second-strike force. Submarine-based missiles do this job quite well. If antisubmarine warfare technology forced military leaders to look upward for a place to station reserve bombing forces, Earth orbit rather than the moon would be the best location. Even if the logic of nuclear deterrence urged that missiles be placed two days' flight time beyond the atmosphere, the right place would be lunar orbit, rather than the surface of the moon. A swarm of missiles in various orbits would be harder to knock out than if they were in fixed launch sites, and the amount of energy required to fire them to their targets would be less. Current international agreements and world public opinion are opposed to an escalation of the arms race into space. So am I, and I see no way such weapons could substantially further the cause of spaceflight.

A considerably more attractive prospect is the idea of an international "Space Patrol" operating in orbit to destroy any ICBM launched by any nation, using laser beams. In his 1958 book *Spaceflight—A Technical Way of Overcoming War,* Eugen Sänger suggested that the problem of destroying ICBM warheads speeding toward their targets at 25,000 kilometers per hour might be solved through the use of "energy rays, which could follow and reach any material body without difficulty, on account of their million times lesser inertia and higher velocity."[16]

Current research on laser weapons indicates that among the problems standing in the way of feasible laser "death rays" are two effects in which gases tend to block or dissipate the beam. In the "thermal blooming" effect, the atmosphere along the path of the ray defocuses it and reduces greatly the amount of energy delivered to the target. ICBM interception from space would not have to deal with this problem because there would be no atmosphere along the path of the laser beam. In the "laser-supported absorption-wave" effect, gas emitted by the target when struck by the laser beam can form a protective shield. Presumably, the hard vacuum of space would encourage this plasma shield to dissipate faster, improving the efficiency of the weapon.[17]

A warhead might be protected against laser fire by a reflective coating, but if it were hit by beams at a variety of wavelengths simultaneously, one at least would be absorbed and degrade the surface enough for the others to take effect. An ablative heat-shield, like that protecting the warhead from reentry damage, would provide some protection against visible wavelength and infrared lasers, but not against the X-ray and gamma-ray lasers that may soon be developed. Perhaps the most difficult technical problem is how the laser beams could be aimed precisely.

To be really effective, the Space Patrol would have to be a rather large and varied force, operating a number of heavy satellites. Such a system would require in-flight maintenance and supply, whether the satellites themselves were manned or unmanned. Clearly a massive orbiting laser defense force would make extensive use of space shuttle transport. This would both reduce space launching costs in general through economies of scale and stimulate the development of more advanced space transport systems.

The Space Patrol would best serve the interests of world peace if it were nonpartisan—designed to shoot down any object that rose above the atmosphere without special permission. To build an international Space Patrol would be advantageous to the United States. It must be entirely nonpartisan, prepared to shoot down even American missiles. Space launches could be announced in advance so the Space Patrol would let them proceed. The introduction of the Space Patrol would temporarily destabilize the international balance of power, giving nations with large conventional forces some advantage. However, the United States need not suffer any loss of security in such a situation. Antiaircraft missiles of both the United States and Russia have just reached the point where systems using them can effectively protect these nations' borders from conventional aircraft and cruise missiles. Thus, the Space Patrol would revolutionize war by making defense superior to offense. This would be a complete reversal of the current situation in which atomic offense can overwhelm defense and peace is maintained only precariously by a balance of terror. Obviously, the Space Patrol idea may turn out to be infeasible for technical or economic reasons. I think it is a good idea; mere words on treaties do not seem to give us world security, and world government is no closer than it was 30 years ago. The only solution may be found in resorting to new technical means to block the technical danger presented by ICBMs. The Space Patrol is the only military project I can think of that is both humanely desirable and would further space technology.

Colonization

Human colonies on other planets would provide a greatly expanded scope for human society and, therefore, are a worthy topic of sociological consideration. However, neither the state of technology nor the environments of the planets encourage me to believe that a wave of colonization will sweep forward in the near future.

Some people still imagine that Earth's population explosion could be solved by shipping excess people off as colonists to new lands in the sky. This idea is ridiculous. Surely we would be colonizing Antarctica or the bottom of the sea already, if room were the issue. Even Alaska has failed to attract many colonists. Indeed, many rural areas in advanced industrial nations have shown a tendency to *lose* population rather than gain it. Colonies would have to be more than mere places to put people; they would have to provide an economic boost for industrial society before investing in them would be reasonable.

Where could colonies be planted? There is a minor movement afoot to build floating cities in Earth orbit, an interesting, if somewhat frivolous, idea.[18] Even if constructed, they would be extensions of our planet, not true colonies. The surfaces of Venus, Jupiter, Saturn, Uranus, and Neptune are completely inhospitable. Smaller bodies in the solar system, asteroids and the satellites with a radius of less than 1000 kilometers, are unlikely to offer the variety of raw materials necessary for the development of an independent economy. The four largest satellites of Jupiter would be attractive places if they were not bathed in the lethal radiation surrounding the planet. There is something to be said for the comets, because spectroscopic analysis suggests they often are composed of a good variety of chemical elements, but the idea of living on or in a comet seems to me more a poetic than a technical idea. We are left with the planet Mercury (which has little to recommend it except high density), the moon (which is at least familiar and nearby), Mars (which remains the best bet), and the satellites of Saturn. Only after a colony has succeeded on one of these bodies should we consider other bodies, such as the satellites of Uranus or Neptune.[19]

Today's technology is not up to the task of planting a really self-sufficient colony, although what has to be done is clear enough that I am reasonably sure we will ultimately be able to inhabit other celestial bodies. Human society is not yet even economically self-sufficient on Earth! Much of the work of civilization has been done for us by the planet itself when it separated many of its materials into conveniently exploitable deposits of ore. Plants and animals, living off the soil and air, produce all our food, many of our chemicals, and some of our basic materials. Although we now pay some price to keep our air and water clean, we at least do not have to manufacture them. As the years pass, we can hope technology will grow more and more able to construct a completely closed-system artificial ecology such as would be necessary on another planet.

Let me suggest two models of the economics of interplanetary colonization. First, the advance of space technology might make the extraction of resources from other planets for use on Earth economically feasible. Colonies would then grow up in a normal way, following the extraterrestrial industries. Investment in the planets would have to pay off in reasonably short periods of time after which further investment would be profitable. I very much doubt that extraterrestrial colonies could ever pay their way through exports to Earth. I certainly do not have good estimates of the cost of mining uranium on the moon, but my assumption is that the terrestrial economy will be moving away from the raw-materials-intensive pattern of the nineteenth and twentieth centuries toward a pattern marked by the transformation of very common materials (e.g., iron, sea water, aluminum) into the physical necessities of life. This book should not be mistaken for a treatise in futurology, but surely no one can believe that the planets will be colonized in the twenty-first century for nineteenth-century reasons. Future technology is more likely to be information-intensive than materials-intensive. We can create all the information we want here at home.

A second model assumes that some noneconomic motive will be the driving force behind colonization and that economic considerations will enter our calculations only in service of the ideal of planting a permanent colony with the potential for slow but steady growth. This model suggests very different technical solutions to the problems of living and working off Earth. In the first model, which is essentially capitalist, the measure of success is a reasonable profit in a fairly short time. In the second model the measure of success is a healthy internal economy designed to fit the local conditions as closely as possible. On bodies like the moon we would be encouraged to develop machines and buildings designed to last tens of thousands of years, while on Earth manufacturers and builders think in terms of 10- to 30-year amortization periods. A properly built lunar city need never wear out. Once put in place, at tremendous expense, further costs would be at an absolute minimum. Dynamic systems that could do their work for centuries without repair are even possible to design. For example, electric power could be generated by thermocouples buried in the lunar soil. One set of thermocouple elements would be buried 1 or 2 centimeters under the surface, where there are great extremes in temperature over the period of a month. Other elements would be buried 2 or 3 meters down where the temperature is constant. The temperature differential would produce electric current, conceivably for 100,000 years without major repair. The thermocouple system would produce far less electric power in its first 10 years than a nuclear generating station of the same weight, but would require no supervision and would produce more power over its total life. Thus, colonies on the moon or other celestial bodies might succeed but are unlikely to be the product of capitalist economies. In their comprehensive introductory text, *Intelligence in the Universe,* MacGowan and Ordway even argue that a biological civilization like our own is far less likely to colonize space than a cybernetic "robot" civilization which would have a much longer time perspective than do we mere mortals.[20]

Hubris

The Apollo program was often compared to the great pyramids of ancient Egypt by opponents of NASA and of large space expenditures. Some felt that the pyramids symbolized useless expenditure on superfluous projects, while others felt that the pyramids were built by coerced labor and therefore were monuments to human cruelty. From another perspective, the pyramids express the pride of the Pharaohs. I have even found a few pro-spaceflight individuals who agreed that

Apollo was akin to the pyramids, but felt that both were admirable achievements. What do we remember the Egyptians for (they commented), except the great pyramids? America will always be remembered for having reached the moon! For many people the space program has been a source of personal and national pride. Conceivably, the space program may gain some support indefinitely from the pride it can generate, and a second wave of manned exploration might be begun in search of new things to be proud of.

As one respondent said, "Although it is frequently pooh-poohed by the typical American with a national guilt complex, he still feels subconsciously proud of living in the technologically most advanced country in the world." Other respondents commented that the space program "presents a strong image of United States leadership" and "increases the respect the rest of the world has for the United States of America." The pioneers and leaders of the Spaceflight Movement seem in great part motivated by pride and the quest for honor. How much pride is healthy is a matter of opinion; overweening pride is often called *hubris,* which in the Classical Greek dramas usually led to the tragic fall of whatever person or people had been seized by such arrogance.

The ideology of the Spaceflight Movement may seem arrogant to some and glorious to others, because it views the entire universe as in some sense *ours.* A member of the Peenemünde team I interviewed in Huntsville said that von Braun and his associates could not believe that the stars exist for no purpose. Because the stars are out there, they must exist to be visited and brought into man's realm. If we cannot reach the stars, either the stars are absurd or we are absurd. Clearly, for von Braun the universe is a place of purpose in which human will is an essential force. Peenemünde engineer Krafft Ehricke expresses great pride-of-species in his "Three Laws of Astronautics":

> I. Nobody and nothing under the natural laws of this universe imposes any limitations on man except man himself.
>
> II. Not only the earth, but the entire solar system, and as much of the universe as he can reach under the laws of nature, are man's rightful field of activity.
>
> III. By expanding through the universe, man fulfills his destiny as an element of life, endowed with the power of reason and the wisdom of the moral law within himself.[21]

Some set of future social conditions may encourage leading individuals or powerful classes in society to invest in a further wave of space exploration. For example, if the entire world were someday organized along lines similar to those of ancient Egypt, the equivalent of the Pharaoh might decide that his pride would be best served by building a city on Mars to bear his name. Alternately, in a future age the world may be dominated by an elite class of technocrats. Particularly if their status were threatened by some rival group, they might seek an outlet for their arrogance in a great project, such as the colonization of Mars or an expedition to a nearby star. Sociologists tend to be uncomfortable with rash speculations about the future structure of society; science fiction writers delight in them. This does not mean that the future is mere fiction or that such kinds of society might not come into being, perhaps even in the near future. There was more than a hint of both the pharaonic and the technocratic in the Apollo program, although I would hesitate to say how much of the pride associated with it was legitimate and how much was hubris.

The descent stage of Spaceship Eagle sits at Tranquility Base, abandoned. On one of its legs there is a plaque which reads: "Here men from the planet Earth first set foot upon the Moon, July 1969, A.D. We came in peace for all mankind." It was signed by Armstrong, Collins, Aldrin, and Nixon. The poet Shelley might have worded it differently: "My name is Richard Nixon, king of kings: Look on my works, ye Mighty, and despair!" Where now is his pride—or ours? We have not yet marked the moon significantly: "Boundless and bare the lone and level sands stretch far away."

Communication with Extraterrestrial Intelligence

A second space revolution may already have begun. For the past 15 years or so, a small network of imaginative and scientifically trained men has been exploring the possibility of making contact with intelligent creatures on other planets, perhaps by radio, and entering into meaningful communication with them. This movement now has a name, CETI, the acronym of Communication with ExtraTerrestrial Intelligence.

Two gradual developments in scientific thinking over the past several decades have set the stage for the CETI movement. First, as the mechanisms and origins of life have been progressively investigated, the feeling has grown that it is not at all unlikely for intelligent life to arise on any reasonably hospitable planet. Second, as theories of the origin of the solar system have evolved, astronomical opinion has shifted from the belief that planets were rare in the universe to the conviction that they are common.

One of the most impressive facts about the CETI movement itself is that its leading members are almost without exception highly respected and productive scientists, usually working in one or another of the most modern branches of Astronomy. For example, the basic arguments in favor of the thesis that life abounds in the universe were expressed in a 1951 book on the subject by the British Astronomer Royal, Sir Harold Spencer Jones.[22] The best-known CETI leader is Carl Sagan, Director of the Laboratory for Planetary Studies at Cornell University, key participant in NASA's Mariner 9 Project (among others), editor of the respected journal *Icarus,* and author of a large number of scientific papers. Of course, this contrasts greatly with the case of the Spaceflight Movement.

Another interesting fact is that the movement is thoroughly international, being particularly strong in the United States, the Soviet Union, and Great Britain. The Spaceflight Movement was equally international around 1930, but the CETI men have the money to exchange far more visits. Perhaps the key book in the field was a collaboration between Sagan and the Russian I. S. Shklovskii.[23] The Russians held a major CETI conference in Soviet Armenia in 1971, attended by 54 scientists: 32 Russians, 18 Americans, and 4 from other nations.[24]

Despite the fact that CETI has become a respected (if peripheral) topic in scientific circles, the drive to achieve contact is being carried on by a social movement, a movement internal to standard institutions but operating outside them to achieve a new goal. It is too early to tell if the CETI movement will resort to truly revolutionary methods of securing social support. One of the tactics followed by the movement has been a continuing attempt to stimulate public interest through articles, books, television appearances, public symposia, and every other conceivable medium. In this, CETI reminds us of the Spaceflight Movement before the German wing began its military detour. One of the most effective means the CETI men have used to generate publicity in both popular and scientific media has been the bold performance of demonstration experiments.

The first of these was Project Ozma, carried out by Frank Drake at the National Radio Astronomy Observatory near Green Bank, West Virginia. In 1960 he pointed the observatory's 85-foot radio telescope at the stars Tau Ceti (note the pun) and Epsilon Eridani, nearby stars sufficiently like our sun that conceivably they had inhabited planets.[25]

There were no scientific results, but Ozma stirred up a good deal of publicity.

Since Ozma, various radio astronomers have occasionally listened for signals from other stars, with similar negative results. When pulsars—natural sources of radio noise that pulsed rapidly—were first discovered they were briefly thought to be artificial. A few imaginative folk temporarily called the pulsars "LGMs," standing for Little Green Men. Drake and Sagan have recently estimated that only 0.1% of the necessary number of stars have been observed to give a 50% chance of success.[26] We cannot tell how serious brief projects like Ozma really are, but they do stimulate public interest.

Most writers on the subject believe that radio wavelengths are most suitable for CETI,[27] but Herbert F. Wischnia has used the Copernicus satellite to observe three stars in hopes of picking up laser beams sent from them.[28]

There have also been a couple of attempts to send messages from Earth to other civilizations. Frank Drake used the recently improved 1000-foot diameter radio telescope at Arecibo, Puerto Rico, to send a brief radio message toward Messier 13, a star cluster.[29] Radio technology has indeed advanced to the point where we can send messages all the way across the galaxy, if only there is someone on the other end with similar technology to receive them. The time scale of any interstellar conversation would be staggering. Our signal to Messier 13 will not arrive for 24,000 years, so if the natives reply immediately upon receiving the radio impulses, we will not get their answer for 48,000 years.

The most publicized CETI event of recent years was a letter that Drake, Sagan, and Linda Sagan (Carl's artistic wife) attached to the *Pioneer 10* Jupiter probe which was launched with sufficient velocity to escape the solar system. The plaque carried a diagram of the solar system showing the point of origin of the probe, a chart of nearby pulsars that might enable anybody who found it to learn when the probe was launched and from what star, and the nude figures of an earthwoman and an earthman saying hello.[30] Copies of the plaque appeared in many newspapers and popular magazines, and a souvenir duplicate of it could be bought. The TRW Corporation, which had built the spacecraft, launched a promotional contest based on the plaque. The contestants were presented with a drawing of it with space to add a picture of the alien being that would receive the message from Earth. There were also cartoonist's balloons over the figures in which the contestant was supposed to write what the three characters were saying to each other.[31] Sagan and Drake admitted that the chance anyone would ever find *Pioneer 10* was infinitesimal, and Sagan apparently did see the plaque primarily as domestic propaganda: "The greater significance of the *Pioneer 10* plaque is not as a message to out there; it is as a message to back here."[32]

Members of the CETI movement have started actively seeking financial support to build huge receiving antennas. Sagan and Drake proclaimed in a recent issue of *Scientific American:*

> [W]e on the earth now possess all the technology necessary for communicating with other civilizations in the depths of space. Indeed, we may now be standing on a threshold about to take the momentous step a planetary society takes but once: first contact with another civilization.[33]

They make it sound as if the main problem were getting together the cash to pay for a radio search of the sky. Sagan and Drake guess that the galaxy contains about 1,000,000 advanced civilizations with which we might communicate, and from that they derive the estimate that the likely distance to the nearest one would be 300 light years.[34] Another major problem is that we do not know at which of several hundred thousand stars to direct our antennas. To have any chance of success, a CETI project would have to scan a large number of stars for a length of time at a variety of wavelengths. Much of the theoretical work being performed by members of the movement is aimed at finding the best guidelines for a CETI search to narrow the odds as much as possible. Bernard M. Oliver has proposed a Project Cyclops to build an orchard of 1500 100-meter antenna dishes at a cost of something like $10,000,000,000 to hunt for extraterrestrial radio signals.[35] This sum is only two fifths of what the Apollo project cost, so some government could conceivably put up the money, if the proper social mechanisms motivated it to do so.

CETI, like spaceflight, is too expensive to be afforded by an amateur movement. In a somewhat naive *Saturday Review* article published in 1972, John Lear urged some "generous person" to come forward and donate $70,000,000 of his own money to build a large radio telescope, like the Arecibo dish, and dedicate it to the quest for CETI.[36] Perhaps one of the world's superrich eccentrics will include a bequest of this magnitude in his will, just as others have financed large optical telescopes. The early chapters of this book report how the Spaceflight Movement failed utterly to attract sufficient independent support, and

I have no reason to expect that the CETI movement will be more lucky.

CETI could hardly be achieved through a military detour, but I see some possibility of a *religious detour*. A number of recent religious movements have had a technological or scientific flavor; Scientology is the best example. All around the country there are little flying saucer clubs and cults that take seriously the proposition that Extraterrestrials have already landed. The Gallup Poll organization found that 54% of a national sample asked in 1974 believed UFOs were "real," while 30% thought they were "people's imagination," and 16% had no opinion.[37] A surprisingly large 11% reported they themselves had seen a UFO. The form of the questions may have biased responses in a positive direction, but the poll does indicate a reservoir of opinion conducive to exploitation by the CETI movement.

A new Scientology-like religion could arise, having as one of its basic premises the prophecy that our world will be saved from its wickedness by flying saucer people if only we could send them an S.O.S. Such a religion could build popular support by drawing on the undoubted ambivalence many people feel toward modern science and on contemporary pessimism about the ability of standard human institutions to solve the world's problems. A Saucerite religion could fund the CETI project directly or use its influence to channel public funds into something like Project Cyclops.

Today there is no connection between the CETI movement and the flying saucer occult subculture. In fact, one of Sagan's hobbies and means of generating publicity is to debunk occult pseudoastronomical notions.[38] The Spaceflight Movement was costly, not only in dollars but also in human lives. It gave us the ICBM, a gift we could have done without. The CETI movement might succeed through a religious detour at the cost of polluting the public mind with quackery and pseudoscience. Revolutions may often be desirable, but they are always dangerous.

There is, of course, some hope that the routine activity of conventional radio astronomers will achieve CETI, without intending to. There have been several false alarms already. In a few years it will be time to write a retrospective book on the CETI movement, but today it is in its infancy.

The consequences of true communication with extraterrestrial intelligence would be stunningly revolutionary. No one can doubt this fact. Because of the distances involved, of course, CETI does not mean an immediate dialogue. But if some civilization is intentionally beaming a message in our direction, it is likely to be in the form of an easily decoded encyclopedia of that civilization's culture and scientific knowledge. The mere fact of receiving such a message would produce a wave of social movements, new religions, and general public hysteria. Our species would want all the information possessed by the Extraterrestrials and would struggle to duplicate their achievements.

Proponents have often suggested that CETI would mean contact not with just one civilization, but with many—perhaps hundreds—already tied together in a galaxy-wide communication network. It is enough to make even the most pedestrian social scientist weep. What are the other civilizations like? What is the nature of galactic society?

CETI might also stimulate the space program. Some solar systems might be much more hospitable to the development of an interplanetary culture than ours. If Mars were larger and Venus rotated more rapidly and had less atmosphere, we might be planning to colonize them even with today's backward technology. If we could contact an interplanetary civilization, we might be stimulated to build one of our own, even though the celestial real estate available to us was less desirable. They might even send us blueprints for the ships and cities we would need.

Conversations across the stars can be carried on only by mature civilizations capable of asking a question, then listening for thousands of years until the answer comes. If our species survives the current period of great danger, we may enter into interstellar dialogues, share scientific knowledge, History, Art, and Philosophy. Eventually we may know the stories of many civilizations and be able to perform comparative quantitative studies of space development. Surely some other planet, not far from here as galactic distances go, became the center of a star-spanning culture through a series of unpredictable events it calls the Spaceflight Revolution.

NOTES

Chapter 1

1. Norman Mailer: *Of a Fire on the Moon* (New York: Signet, 1971), p. 69.
2. Neil Armstrong et al.: *First on the Moon* (Boston: Little, Brown, 1970), p. 363.
3. Lincoln P. Bloomfield (ed.): *Outer Space–Prospects for Man and Society* (New York: Praeger, 1968), *Introduction,* p. 3.
4. J. G. Crowther: "Beyond the Pillars of Hercules," *New Scientist,* 17 July 1969, p. 144.
5. V. Gordon Childe: *Man Makes Himself* (New York: Mentor, 1951), p. 59.
6. *Ibid.,* p. 63.
7. Freeman Dyson: "Human Consequences of the Exploration of Space." In Eugene Rabinowitch and Richard S. Lewis (eds.): *Man on the Moon* (New York: Harper & Row, 1970), pp. 26–27.
8. G. Harry Stine: "The Third Industrial Revolution," *Spaceflight,* September 1974, p. 328.
9. John Jewkes, David Sawers, and Richard Stillerman: *The Sources of Invention* (New York: Norton, 1969), p. 76.
10. *Ibid.,* p. 77.
11. *Spaceflight,* October 1969, p. 356.
12. Arthur C. Clarke: "The Challenge of the Spaceship," *Journal of the British Interplanetary Society,* December 1946, p. 68.
13. *Journal of the British Interplanetary Society,* March 1950, p. 57.
14. Jonathan Norton Leonard: *Flight into Space* (New York: Random House, 1957), p. 80.
15. Harold Leland Goodwin: *The Images of Space* (New York: Holt, Rinehart & Winston, 1965), p. 26.
16. *Ibid.,* p. 27.
17. Raymond A. Bauer: *Second-Order Consequences* (Cambridge, Mass.: M.I.T. Press, 1969), p. 84.
18. *Ibid.,* p. 85.
19. Hugo Young, Bryan Silcock, and Peter Dunn: *Journey to Tranquility* (Garden City, N.Y.: Doubleday, 1970), p. 162.

20. Goodwin: *Op. Cit.*, p. 139.
21. *Science,* 24 July 1964, p. 368.
22. Bernd Ruland: *Wernher von Braun–Mein Leben für die Raumfahrt* (Offenburg, W. Germany: Burda, 1969), p. 223.
23. Young et al.: *Op. Cit.*, p. 19.
24. Thomas S. Kuhn: *The Structure of Scientific Revolutions* (Chicago: University of Chicago Press, 1962), p. 10.
25. *Ibid.*, p. 91.
26. S. C. Gilfillan: *The Sociology of Invention* (Cambridge, Mass.: M.I.T. Press, 1963), p. 10.
27. James M. Utterback: "Innovation in Industry and the Diffusion of Technology," *Science,* 15 February 1974, p. 621.
28. Jacob Schmookler: *Invention and Economic Growth* (Cambridge, Mass.: Harvard University Press, 1966), p. 206.
29. *Ibid.*, p. 69.
30. *Ibid.*, p. 194.
31. Leslie E. Simon: *Secret Weapons of the Third Reich* (Old Greenwich, Conn.: WE, Inc., 1971), p. 91.
32. I. Essers: *Max Valier, Ein Vorkämpfer der Weltraumfahrt* (Düsseldorf: Verein Deutscher Ingenieure, 1968), p. 174.
33. Eugen Sänger: *Raumfahrt-Technische Überwindung des Krieges* (Hamburg: Rowohlt, 1958).

Chapter 2

1. Neil J. Smelser: *Theory of Collective Behavior* (New York: Free Press, 1962), p. 297.
2. Edward B. Roberts: "Enterpreneurship and Technology," in William H. Gruber and Donald G. Marquis (eds.): *Factors in the Transfer of Technology* (Cambridge, Mass.: M.I.T. Press, 1969), pp. 219–237, especially 236.
3. *Ibid.*, p. 223.
4. Wernher von Braun: *The Mars Project* (Urbana: University of Illinois Press, 1962), p. 1.
5. Martin Gardner: *Fads and Fallacies in the Name of Science* (New York: Dover, 1957), p. 11.
6. Edward M. Lemert: "Paranoia and the Dynamics of Exclusion," in *Human Deviance, Social Problems, and Social Control* (Englewood Cliffs, N.J.: Prentice-Hall, 1972).
7. Thomas J. Scheff: *Being Mentally Ill* (Chicago: Aldine, 1966); Erving Goffman: *Asylums* (Garden City, N.Y.: Doubleday, Anchor, 1961); D. L. Rosenhan: "On Being Sane in Insane Places," *Science,* 19 January 1973, pp. 250–258, but, for contrary opinions, see: Madeline Karmel: "Total Institution and Self-Mortification," *Journal of Health and Social Behavior,* 1969, pp. 134–140; Walter R. Gove: "Societal Reaction as an Explanation of Mental Illness: An Evaluation," *American Sociological Review,* October 1970, pp. 873–884.
8. N. A. Rynin: *Interplanetary Flight and Communication* (Jerusalem: Israel Program for Scientific Translations, 1971), Volume 4, p. 37.

9. *Ibid.*, p. 38.
10. *Ibid.*, Willy Ley: *Rockets, Missiles and Men in Space* (New York: Signet, 1969), pp. 113–115.
11. Evgeny Riabchikov: *Russians in Space* (Garden City, N.Y.: Doubleday, 1971), p. 88.
12. Rynin: *Op. Cit.*, Volume 4, p. 39.
13. *Ibid.*, p. 41.
14. *Ibid.*, p. 33.
15. Ley: *Op. Cit.*, pp. 113–115.
16. Robert K. Merton: *The Sociology of Science* (Chicago: University of Chicago Press, 1973), pp. 286–287.
17. *Ibid.*, p. 289.
18. Heinz Gartmann: *The Men behind the Space Rockets* (New York: McKay, 1956), p. 13.
19. *Ibid.*, p. 15.
20. Ley: *Op. Cit.*, p. 123.
21. Ley: *Op. Cit.*
22. Rynin: *Op. Cit.*, Volume 4, p. 162.
23. Essers: *Op. Cit.*, p. 122.
24. Ley: *Op. Cit.*, p. 122.
25. Hermann Oberth: *Die Rakete zu den Planetenräumen* (München: Oldenbourg, 1923), pp. 90–92.
26. Leonid Vladimirov: *The Russian Space Bluff* (New York: Dial, 1973), p. 27.
27. Rynin: *Op. Cit.*, Volume 7, p. 2.
28. Riabchikov: *Op. Cit.*, p. 91.
29. A. Kosmodemyansky: *Konstantin Tsiolkovsky–His Life and Work* (Moscow: Foreign Languages Publishing House, 1956), p. 9.
30. Rynin: *Op. Cit.*, Volume 7, p. 29.
31. Kosmodemyansky: *Op. Cit.*, p. 9.
32. Konstantin Tsiolkovsky: *The Call of the Cosmos* (Moscow: Foreign Languages Publishing House, 1960), p. 80.
33. *Ibid.*
34. Riabchikov: *Op. Cit.*, p. 98.
35. Arthur J. Deikman: "Deautomatization and the Mystical Experience" in Charles Tart (ed.): *Altered States of Consciousness* (New York: Wiley, 1969), pp. 23–43; A. C. Bhaktivedanta Swami: *Easy Journey to Other Planets* (Boston: Iskon, 1970); Emanuel Swedenborg: *The Earths in our Solar System, Which Are Called Planets, Etc.* (Boston: New Church Union, 1950 [1758]).
36. Tsiolkovsky: *Call of the Cosmos*, p. 10.
37. Rynin: *Op. Cit.*, Volume 7,p. 2.
38. *Ibid.*, p. 3.
39. Ley: *Op. Cit.*, p. 125.
40. Georg Simmel: *The Sociology of Georg Simmel* (New York: Free Press, 1950), p. 404.
41. Jewkes et al.: *Op. Cit.*, p. 96.
42. Gilfillan: *Op. Cit.*, pp. 11–12.
43. Rynin: *Op. Cit.*, Volume 7, p. 3.

44. Kosmodemyansky: *Op. Cit.*, p. 11.
45. Abraham H. Maslow: *Religions, Values and Peak-Experiences* (New York: Viking, 1970).
46. Rynin: *Op. Cit.*, Volume 7, p. 3.
47. Kosmodemyansky: *Op. Cit.*, p. 12.
48. John W. Campbell, Jr.: "The Space-Drive Problem," *Astounding Science Fiction*, June 1960, pp. 83–106.
49. *Ibid.*, p. 99.
50. *Ibid.*
51. *Ibid.*, p. 87.
52. *Ibid.*, p. 91.
53. Robert H. Goddard: *The Autobiography of Robert Hutchings Goddard* (Worcester, Mass.: St. Onge, 1966), pp. 31–33.
54. Milton Lehman: *This High Man* (New York: Pyramid, 1963), p. 33.
55. *Ibid.*, p. 34.
56. Goddard: *Op. Cit.*, pp. 33–34.
57. Lehman: *Op. Cit.*, p. 38.
58. *Ibid.*, pp. 35, 36.
59. *Ibid.*, p. 57.
60. Esther C. Goddard and Edward Pendray (eds.): *The Papers of Robert H. Goddard* (New York: McGraw-Hill, 1970), Volume I, p. 27.
61. Lehman: *Op. Cit.*, p. 75.
62. *Ibid.*, p. 76.
63. Esther C. Goddard: *Op. Cit.*, p. 22.
64. G. Edward Pendray: "The Man who Ushered in the Space Age," introduction to Robert H. Goddard: *Rocket Development* (Englewood Cliffs, N.J.: Prentice-Hall, 1961), p. xiv.
65. Robert H. Goddard: "A Method of Reaching Extreme Altitudes," in Esther C. Goddard: *Op. Cit.*, pp. 337–406.
66. Ley: *Op. Cit.*, pp. 160, 237.
67. Robert Goddard: *Rocket Development.*
68. *Ibid.*, p. xix.
69. Eugene Emme: *A History of Space Flight* (New York: Holt, Rinehart & Winston, 1965), p. 91.
70. *New York Times*, 13 January 1920: "Editorial on Goddard," reprinted in Arthur C. Clarke: *The Coming of the Space Age* (New York: Meredith, 1967), p. 66.
71. Esther Goddard: *Op. Cit.*, p. 32.
72. Lehman: *Op. Cit.*, p. 223.
73. Hermann Oberth: "Autobiography," in Clarke: *The Coming of the Space Age*, p. 115.
74. Hans Hartl: *Hermann Oberth–Vorkämpfer der Weltraumfahrt* (Hannover, Germany: Oppermann, 1958), p. 59.
75. Wernher von Braun and Frederick I. Ordway: *History of Rocketry and Space Travel* (New York: Crowell, revised edition, ca. 1972), p. 58.
76. Lehman: *Op. Cit.*, p. 57.

77. Oberth: "Autobiography," in Clarke: *Op. Cit.,* p. 118.
78. Oberth: *Die Rakete zu den Planetenräumen.*
79. Arthur C. Clarke: *The Exploration of Space* (Greenwich, Conn.: Fawcett, 1964), p. 27.
80. Hartl: *Op. Cit.,* pp. 68–69.
81. Hermann Oberth: *Stoff und Leben* (Remagen, Germany: Reichl, 1959).
82. Essers: *Op. Cit.,* pp. 87–88.
83. Oberth: *Stoff und Leben,* p. 15.
84. *Ibid.,* p. 172.
85. Hermann Oberth: *Man into Space* (New York: Harper, 1957), pp. 166–167.
86. Magnus Freiherr von Braun: *Weg durch vier Zeitepochen* (Limburg an der Lahn Germany: Starke, 1965).
87. Bernd Ruland: *Wernher von Braun–Mein Leben für die Raumfahrt* (Offenburg, Germany: Burda, 1969), p. 49.
88. John C. Goodrum: *Wernher von Braun, Space Pioneer* (Strode, 1969), p. 15.
89. Ruland: *Op. Cit.,* p. 51.
90. Goodrum: *Op. Cit.,* p. 19.
91. G. Edward Pendray: *The Coming Age of Rocket Power* (New York: Harper, 1945), pp. 110–111.
92. Evgeny Riabchikov: *Russians in Space* (Garden City, N.Y.: Doubleday, 1971), p. 105.
93. Michael Stoiko: *Soviet Rocketry* (New York: Holt, Rinehart & Winston, 1970), p. 47.
94. *Ibid.,* p. 65.
95. K. W. Gatland: "Rocketry in the Soviet Union," *Science Journal,* November 1968, p. 73.
96. Riabchikov: *Op. Cit.,* p. 139.
97. Albert Parry: *Russia's Rockets and Missiles* (Garden City, N.Y.: Doubleday, 1960), p. 164.
98. Leonid Vladimirov: *The Russian Space Bluff* (New York: Dial Press, 1973).
99. Riabchikov: *Op. Cit.,* p. 139.
100. Vladimirov: *Op. Cit.,* p. 24.
101. "Talking about Korolyov," *New Scientist,* 23 May 1974, p. 496.
102. Sources for biographical information were as follows:

    Wernher von Braun—Ruland: *Op. Cit.*; Goodrum: *Op. Cit.*; Erik Bergaust: *Reaching for the Stars* (Garden City, N.Y.: Doubleday, 1960); Shirley Thomas: *Men of Space* (Philadelphia: Chilton, 1961–1968, 8 volumes), Volume 1, pp. 133–156; *Wer ist Wer?,* XVII, (Frankfurt: Societäts-Verlag, 1973), p. 117.

    Arthur C. Clarke—Thomas: *Op. Cit.,* Volume 8, pp. 14–16; Sam Moskowitz: *Seekers of Tomorrow* (Westport, Conn.: Hyperion, 1974), pp. 374–391; *The Author's and Writer's Who's Who* (London: Burke's Peerage, 1971), p. 153.

    Walter Dornberger—Thomas: *Op. Cit.,* Volume 2, pp. 44–63; *Wer ist Wer?,* XVI, (Berlin: Arani, 1970), p. 220.

    Krafft A. Ehricke—Thomas: *Op. Cit.,* Volume 1, pp. 1–22; Marjorie Dent Candee (ed.): *Current Biography,* 1958, pp. 130–132.

Robert H. Goddard—Thomas: *Op. Cit.,* Volume 1, pp. 23–46; Lehman: *Op. Cit.*; Robert H. Goddard: *Autobiography* (Worcester, Mass.: St. Onge, 1966).

Hermann Ganswindt—Ley: *Op. Cit.,* pp. 115–124; Gartmann: *Op. Cit.*

Nikolai Kibal'chich—Ley: *Op. Cit.,* pp. 113–115; Rynin: *Op. Cit.,* Volume 4, pp. 36–43.

Sergei Korolyov—Riabchikov: *Op. Cit.,* pp. 86–90.

Willy Ley—Donald H. Tuck: *The Encyclopedia of Science Fiction and Fantasy* (Chicago: Advent, 1974), Volume 1, pp. 274–276; Rynin: *Op. Cit.,* Volume 8, p. 312; *The Author's and Writer's Who's Who* (London: Burke's Peerage, 1971), p. 493; Candee: *Current Biography,* 1953, pp. 356–359.

Hermann Oberth—Hartl: *Op. Cit.*; Candee: *Current Biography,* 1957, pp. 416–418; *Wer ist Wer?,* XVII (Frankfurt: Societäts-Verlag, 1973, p. 788; *Who's Who in Germany* (Munich: Intercontinental [Oldenbourg], 1964) 3rd edition, pp. 1253–1254.

Eugen Sänger—Irene Sänger-Bredt: "The Silver Bird Story," *Spaceflight,* May 1973, pp. 161–181; Thomas: *Op. Cit.,* Volume 8, pp. 34–39; *Wer ist Wer?,* XIV (Berlin-Grunewald: Arani, 1962), p. 1305; *Who's Who in Germany* (Munich: Intercontinental [Oldenbourg], 1964) 3rd edition, p. 1452.

Friderikh Tsander—Nicholas Daniloff: *The Kremlin and the Cosmos* (New York: Knopf, 1972), pp. 20–25.

K. E. Tsiolkovsky—Rynin: *Op. Cit.,* Volume 7; Kosmodemyansky: *Op. Cit.*; Ley: *Op. Cit.,* pp. 124–129; Thomas: *Op. Cit.,* Volume 1, pp. 90–110; Arthur C. Clarke: *The Coming of the Space Age* (New York: Meredith, 1967), pp. 100–104.

Max Valier—Essers: *Op. Cit.*

103. Personal communication.

104. Daniel Bell: "Notes on the Post-Industrial Society," in Jack D. Douglas: *The Technological Threat* (Englewood Cliffs, N.J.: Prentice-Hall, 1971), p. 13.

105. Percival Lowell: *Mars as an Abode of Life* (New York: Macmillan, 1908).

106. Essers: *Op. Cit.,* p. 41ff.; Willy Ley: "Pseudoscience in Naziland," *Astounding Science Fiction,* May 1947, pp. 90–98.

107. Poul Anderson: *Tau Zero* (New York: Lancer, 1970), p. 57.

108. A number of lay people, of course, believed spaceflight was easy to accomplish. I do not mean my blanket statement to stand instead of a thorough study of attitudes toward spaceflight or other futuristic possibilities that were in the public mind during this historical period. The biographies of the early pioneers are filled with incidents in which the rocket man was rebuffed, denounced, refused honor, and so forth, on the basis that spaceflight was impractical and uninteresting.

109. Emme: *Op. Cit.,* pp. 83, 75.

110. *Ibid.,* p. 78.

111. Goddard: *Autobiography,* pp. 30, 31.

112. Lehman: *Op. Cit.,* pp. 174, 228.

113. *Ibid.,* pp. 174, 31.

114. Rynin: *Op. Cit.,* Volume 7, p. 29.

115. Kurd Lasswitz: *Two Planets* (Carbondale: Southern Illinois University Press, 1971).

116. Von Braun and Ordway: *Op. Cit.,* p. 56.

Chapter 3

1. Hans Hartl: *Hermann Oberth–Vorkämpfer der Weltraumfahrt* (Hannover: Oppermann, 1958), p. 104.
2. Willy Ley: *Rockets, Missiles, and Men in Space* (New York: Signet, 1969), p. 150; John Baxter: *Science Fiction in the Cinema* (New York: Paperback Library, 1970), p. 35.
3. Quoted by John McPhee: *The Curve of Binding Energy* (New York: Ballantine, 1975), pp. 129–130.
4. Ley: *Op. Cit.*, p. 190.
5. Walter R. Dornberger: "European Rocketry after World War I," in L. J. Carter (ed.): *Realities of Space Travel* (London: Putnam, 1957), p. 382.
6. Ernst Klee and Otto Merk: *Damals in Peenemünde* (Oldenburg, West Germany: Stalling, 1963), p. 21.
7. I. Essers: *Max Valier–Ein Vorkämpfer der Weltraumfahrt* (Düsseldorf: Verein Deutscher Ingenieure, 1968), p. 184.
8. Ley: *Op. Cit.*, pp. 185–188; Martin Gardner: *Fads and Fallacies in the Name of Science* (New York: Dover, 1957), pp. 19–27.
9. Ley: *Op. Cit.*, p. 187.
10. Herbert Rosinski: "The Remaking of the German Army," in Gordon B. Turner (ed.): *A History of Military Affairs in Western Society since the Eighteenth Century* (Ann Arbor, Mich.: Edwards Brothers, 1952), p. 675.
11. John Killen writes:

    [T]he calculating brain and guiding hands of General von Seeckt were secretly at work behind the outwardly innocent facade of Lufthansa, manipulating his chosen men into the important positions in German civil aviation, and ensuring the future development of military projects in a civilian disguise that successfully blinded the world to the deadly underlying purpose.

    John Killen: *A History of the Luftwaffe* (New York: Berkley Medalion, 1970), pp. 50–51.)
12. Paul Carell: *Unternehmen Barbarossa* (Frankfurt: Ullstein, 1968), Volume I, p. 184.
13. Ernst Heinkel: *Stürmisches Leben* (Stuttgart, West Germany: Mundus, 1953), pp. 134–135.
14. Ley: *Op. Cit.*, p. 71.
15. *Ibid.*, p. 195.
16. Shirley Thomas: *Men of Space* (Philadelphia: Chilton, 1961), Volume II, p. 48.
17. Ley: *Op. Cit.*, p. 183.
18. *Ibid.*, p. 230.
19. Walter Dornberger: *V2–Der Schuss ins Weltall* (Esslingen am Neckar, W. Germany: Bechtle, 1952), p. 26.
20. *Ibid.*
21. Thomas: *Op. Cit.*, Volume II, p. 49.

22. *Ibid.*
23. Essers: *Op. Cit.,* pp. 242–243.
24. Thomas: *Op. Cit.,* Volume II, p. 50.
25. Dornberger: *V2,* pp. 26–27.
26. Bernd Ruland: *Wernher von Braun–Mein Leben für die Raumfahrt* (Offenburg, W. Germany: Burda, 1969), p. 74. (My translation.)
27. Thomas: *Op. Cit.,* Volume II, p. 50.
28. Willy Ley: "Count von Braun," *Journal of the British Interplanetary Society,* June 1947, p. 155.
29. Heinz Gartmann: *The Men behind the Space Rockets* (New York: McKay, 1956), p. 138.
30. Dornberger: *V2,* p. 45.
31. Ruland: *Op. Cit.,* p. 90.
32. Heinkel: *Op. Cit.,* pp. 472–497.
33. *Ibid.,* pp. 223–228.
34. James C. Sparks: *Winged Rocketry* (New York: Dodd, Mead, 1968), pp. 16–31.
35. W. H. Tantum and E. J. Hoffschmidt: *The Rise and Fall of the German Air Force* (Old Greenwich, Conn.: WE, Inc., 1969), p. 67.
36. Wernher von Braun: "From Small Beginnings...," in Kenneth W. Gatland (ed.): *Project Satellite* (New York: British Book Centre, 1958), p. 32.
37. Dornberger: *V2,* p. 48; Ley: *Rockets,* pp. 229, 242; Ruland: *Op. Cit.,* p. 96; Klee and Merk: *Op. Cit.,* p. 22.
38. Leslie E. Simon: *Secret Weapons of the Third Reich* (Old Greenwich, Conn.: WE, Inc., 1971) (1945 military document), p. 33.
39. Ruland: *Op. Cit.,* p. 97.
40. *The Story of Peenemünde,* 1945, p. 117, U.S. Army mimeographed documents. This collection of valuable documents has never been published but can be located in some libraries in mimeographed or other special form. It carries catalogue number KG 6114 in Harvard's library system.
41. *Ibid.,* p. 98.
42. Erik Bergaust: *Reaching for the Stars* (Garden City, N.Y.: Doubleday, 1960), p. 85.
43. *Ibid.,* p. 86.
44. *The Story of Peenemünde,* p. 117.
45. Von Braun: *Op. Cit.,* p. 33.
46. Dornberger: *V2,* p. 21.
47. Dieter K. Huzel: *Peenemünde to Canaveral* (Englewood Cliffs, N.J.: Prentice-Hall, 1962), p. 22.
48. Bergaust: *Op. Cit.,* p. 85; Ruland: *Op. Cit.,* p. 123.
49. Huzel: *Op. Cit.,* p. 33.
50. Clarence G. Lasby: *Project Paperclip–German Scientists and the Cold War* (New York: Atheneum, 1971), p. 14.
51. Ley: *Op. Cit.,* p. 261.
52. Lasby: *Op. Cit.,* p. 269.
53. Bergaust: *Op. Cit.,* p. 77.

54. David Irving: *The Mare's Nest* (Boston: Little, Brown, 1964), p. 26.
55. Dornberger: *V2,* p. 112.
56. Bernard Barber: *Science and the Social Order* (New York: Collier, 1970 [1952]), p. 101.
57. *Ibid.,* p. 113.
58. Simon: *Op. Cit.,* p. 162.
59. Helmut Heiber (ed.): *Goebbels-Reden* (Düsseldorf: Droste, 1972), Volume II, p. 277.
60. Helmut Heiber: *Goebbels* (New York: Hawthorn, 1972), pp. 300–301.
61. Ruland: *Op. Cit.,* p. 149.
62. Heiber: *Goebbels-Reden,* pp. 356–357.
63. Gartmann: *Op. Cit.*; Edwart T. Maloney and Uwe Feist: *Messerschmitt 163* (Fallbrook, Calif.: Aero, 1968).
64. Irene Sänger-Bredt: "The Silver Bird Story," *Spaceflight,* May 1973, Vol. 15, No. 5, pp. 166–181; Ley: *Op. Cit.,* p. 512.
65. Thomas: *Op. Cit.,* Volume VIII, p. 37.
66. Sänger-Bredt: *Op. Cit.,* pp. 166, 172.
67. Eugen Sänger and Irene Sänger-Bredt: *A Rocket Drive for Long Range Bombers* (Santa Barbara, Calif.: Cornog, 1952 [1944]).
68. Eugen Sänger: *Raumfahrt–Technische Überwindung des Krieges* (Hamburg: Rowohlt, 1958).
69. Ruland: *Op. Cit.,* p. 239.
70. *Ibid.,* p. 249; Huzel: *Op. Cit.,* p. 139.
71. Wernher von Braun and Frederick I. Ordway: *History of Rocketry and Space Travel* (New York: Crowell, ca. 1970), pp. 114–116; James McGovern: *Crossbow and Overcast* (London: Hutchinson, 1965), pp. 109–112.
72. Huzel: *Op. Cit.,* pp. 151–165.
73. Von Braun and Ordway: *Op. Cit.,* p. 116.
74. Lasby: *Op. Cit.,* p. 125.
75. Ruland: *Op. Cit.,* p. 275.
76. McGovern: *Op. Cit.;* Lasby: *Op. Cit.*
77. *The Story of Peenemünde,* p. 208. This article was recently reprinted. Wernher von Braun: "An Historical Essay," *Spaceflight,* Vol. 14, No. 11, November 1972, pp. 409–412.
78. Klee and Merk: *Op. Cit.,* p. 110.
79. Lasby: *Op. Cit.,* p. 5.
80. *Ibid.,* p. 288.
81. Ruland: *Op. Cit.,* p. 267.
82. *Ibid.,* p. 298.
83. *Ibid.,* p. 284.
84. McGovern: *Op. Cit.,* p. 186.
85. Lasby: *Op. Cit.,* p. 168.
86. *Ibid.,* p. 206.
87. Albert Parry: *Russia's Rockets and Missiles* (Garden City, N.Y.: Doubleday, 1960), p. 121.

88. *Ibid.*, p. 122; K. W. Gatland: "Rocketry in the Soviet Union," *Science Journal*, November 1968, p. 75.
89. Parry: *Op. Cit.*, p. 124.
90. Gatland: *Op. Cit.*, p. 75.
91. Parry: *Op. Cit.*, p. 123.
92. Irmgard Gröttrup: *Die Besessenen und die Mächtigen* (Stuttgart: Steingrüben, 1958).
93. Ley: *Op. Cit.*, p. 266; Gröttrup: *Op. Cit.*, p. 179.
94. Dornberger: *V2* (American Edition) (New York: Ballantine, 1954), p. 179.
95. Ruland: *Op. Cit.*, pp. 187–202.
96. See: Jerry Grey and Vivian Grey: *Space Flight Report* (New York: Basic Books), p. 180.
97. G. A. Tokaev: *Stalin Means War* (London: Weidenfeld and Nicolson, 1951), p. 98.
98. *Ibid.*, pp. 99–103.
99. *Ibid.*, p. 105.
100. *Ibid.*
101. *Ibid.*, p. 122.
102. *Ibid.*, p. 115.
103. *Ibid.*
104. *Ibid.*, pp. 136, 141.
105. Lasby: *Op. Cit.*, p. 9.
106. *Ibid.*, p. 186.
107. John L. Chapman: *Atlas–The Story of a Missile* (New York: Harper, 1960), p. 23.
108. Wernher von Braun: "The Redstone, Jupiter, and Juno," *Technology and Culture*, Fall 1963, p. 453.
109. Richard S. Lewis: *Appointment on the Moon* (New York: Ballantine, 1969), p. 35.
110. Ley: *Op. Cit.*, p. 373.
111. Eugene M. Emme: *A History of Space Flight* (New York: Holt, Rinehart & Winston, 1965), p. 14.
112. Cornelius Ryan (ed.): *Across the Space Frontier* (New York: Viking, 1952).
113. Constance McLaughlin Green and Milton Lomask: *Vanguard: A History* (Washington, D.C.: Smithsonian Institution, 1971), p. 17.
114. Green and Lomask: *Op. Cit.*
115. *Ibid.*
116. *Ibid.*, p. 48.
117. Lewis: *Op. Cit.*, p. 46.
118. *Ibid.*, p. 47.
119. Green and Lomask: *Op. Cit.*, p. 203.
120. Vernon Van Dyke: *Pride and Power–The Rationale of the Space Program* (Urbana: University of Illinois Press, 1964), p. 140.
121. Julian Hartt: *The Mighty Thor–Missile in Readiness* (New York: Duell, Sloan and Pearce, 1961).
122. *Ibid.*, p. 5.
123. *Ibid.*, p. 45.
124. Edgar M. Bottome: *The Missile Gap* (Cranbury, N.J.: Fairleigh Dickinson Press, 1971).

125. Jack Ruina: "SALT in a MAD World," *New York Times Magazine*, 30 June 1974, p. 8.
126. John B. Medaris: *Countdown for Decision* (New York: Putnam, 1960), pp. 54, 57.
127. *Ibid.*, p. 69.
128. *Ibid*, p. 125.
129. John Noble Wilford: *We Reach the Moon* (New York: Norton, 1971), p. 75.
130. Von Braun: "The Redstone, Jupiter and Juno," p. 464.
131. Erik Bergaust and Seabrook Hull: *Rocket to the Moon* (Princeton, N.J.: van Nostrand, 1958), pp. 87–118.
132. Donald Cox and Michael Stoiko: *Spacepower* (Philadelphia: Winston, 1958).
133. Medaris: *Op. Cit.*, p. 135.
134. Personal communication with Konrad Dannenberg, former Peenemünde specialist currently at Huntsville.
135. Philip J. Klass: *Secret Sentries in Space* (New York: Random House, 1971); G. E. Perry: "Cosmos Coverage of the Indo-Pakistani War," *Spaceflight*, September 1972, p. 350.
136. Medaris: *Op. Cit.*, p. 136.
137. David Halberstam: *The Best and the Brightest* (Greenwich, Conn.: Fawcett, 1973).
138. John M. Logsdon: *The Decision to Go to the Moon* (Cambridge, Mass.: M.I.T. Press, 1970), p. 115.
139. Van Dyke: *Op. Cit.*
140. Logsdon: *Op. Cit.*
141. Robert Gillette: "The Soviet Space Program," *Science*, 18 February 1972, p. 733; Robert Hotz: "Soviet Space Problems," *Aviation Week*, 7 May 1973, p. 9.
142. *The Economist:* "The Crew of the Salyut Are Dead," 3–9 July 1971, pp. 157–17; *Der Spiegel:* "Das Ende von Sojus 11," 5 July 1971, pp. 102–110.
143. Ley: *Op. Cit.*, p. 539.
144. Frank Gibney and George J. Feldman: *The Reluctant Space-Farers* (New York: New American Library, 1965), p. 44.
145. Richard Nixon: "The Space Shuttle," *Spaceflight*, April 1972, p. 122.
146. "Columbium RCS Engines Tested," *Aviation Week*, 28 Jan., 1974, p. 44.

Chapter 4

1. Willy Ley: *Rockets, Missiles, and Men in Space* (New York: Signet, 1969), p. 245
2. *Ibid.*, p. 606.
3. David Irving: *The Mare's Nest* (Boston: Little, Brown, 1964) p. 56.
4. G. Edward Pendray: *The Coming Age of Rocket Power* (New York: Harper, 1945), p. 18.
5. J. L. Nayler: *A Dictionary of Astronautics* (New York: Hart, 1964), p. 86.
6. Dieter K. Huzel and David H. Huang: *Design of Liquid Propellant Rocket Engines* (Washington, D. C.: NASA, 1971); see also: Richard T. Holzmann: *Chemical Rockets* (New York: Marcel Dekker, 1969) for a thorough discussion of all these issues.

7. John D. Clark: *Ignition!* (New Brunswick, N. J.: Rutgers University Press, 1972), p. 97.
8. Joseph E. McGolrick: *Launch Vehicle Estimating Factors for Advance Mission Planning* (Washington, D. C.: NASA, 1972), p. 11-1.
9. Nayler: *Op. Cit.,* p. 265.
10. Ley: *Op. Cit.,* pp. 88, 343.
11. Irving: *Op. Cit.,* p. 150.
12. *Ibid.,* p. 44.
13. *Ibid.,* p. 150.
14. *The McGraw-Hill Encyclopedia of Space* (New York: McGraw-Hill, 1967), p. 39.
15. Harvey M. Sapolsky: *The Polaris System Development* (Cambridge, Mass.: Harvard University Press, 1972), p. 8.
16. Ernst Klee and Otto Merk: *Damals in Peenemünde* (Oldenburg, Germany: Stalling, 1963), pp. 104-108.
17. Ley: *Op. Cit.,* p. 293.
18. *Ibid.,* p. 315.
19. *Ibid.,* p. 401.
20. *Ibid.,* p. 411.
21. Erwin J. Bulban: "Fuel Storage Key to Lance Production," *Aviation Week,* March 26, 1973, pp. 52-55.
22. John Cookson and Judith Nottingham: *A Survey of Chemical and Biological Warfare* (New York: Monthly Review Press, 1969), p. 220
23. Bulban: *Op. Cit.,* p. 53.
24. Michael L. Yaffee: "Army Seeks to Modularize Lance Missile," *Aviation Week,* 12 November 1973, pp. 19–20.
25. Wernher von Braun and Frederick I. Ordway III: *History of Rocketry and Space Travel* (New York: Crowell, ca. 1970), revised edition, p. 139.
26. Michael L. Yaffe: "USAF Reconfigures Minuteman ICBMs," *Aviation Week,* 5 November 1973, pp 54–55.
27. Albert Speer: *Erinnerungen* (Berlin: Propyläen, 1969), p. 242.
28. Albert Parry: *Russia's Rockets and Missiles* (Garden City, N.Y.: Doubleday, 1960), p. 114.
29. General Leslie E. Simon: *Secret Weapons of the Third Reich* (Old Greenwich, Conn.: WE, Inc., 1971), p. 130.
30. Irving: *Op. Cit.,* p. 295.
31. Kosmodemyansky: *Op. Cit.,* p. 9.
32. *Ibid.*
33. *Ibid.*
34. Rudolf Lusar: *Die deutschen Waffen und Geheimwaffen des 2. Weltkrieges und ihre Weiterentwicklung* (München: Lehmanns, 1971), p. 196.
35. A proximity fuse which would have circumvented this difficulty was under development, but it was not ready by the end of the war. See Walter Dornberger: "The German V-2," *Technology and Culture,* Fall 1963, pp. 393–409, especially p. 408.
36. Ley: *Op. Cit.,* p. 271.
37. Rynin: *Op. Cit.,* Volume 7, p. 2.
38. *Ibid.,* pp. 271, 269.

39. Irving: *Op. Cit.,* p. 314.
40. Speer: *Op. Cit.,* p. 571.
41. Charles J. Hitch and Roland N. McKean: *The Economics of Defense in the Nuclear Age* (New York: Atheneum, 1965), p. 23.
42. Speer: *Op. Cit.,* p. 329.
43. Walter Dornberger: *V-2* (New York: Ballantine, 1954), p. 236.
44. Arthur M. Squires: "Clean Fuels from Coal Gasification," *Science,* 19 April 1974, pp. 340–346.
45. Irving: *Op. Cit.,* pp. 49, 82.
46. Ley: *Op. Cit.,* pp. 257–258.
47. Dornberger: *V-2* (Ballantine edition), p. 236; Speer: *Op. Cit.,* pp. 361–362.
48. Bernd Ruland: *Wernher von Braun–Mein Leben für die Raumfahrt* (Offenburg, Germany: Burda, 1969), p. 257.
49. James McGovern: *Crossbow and Overcast* (London: Hutchinson, 1964), p. 134.
50. Speer: *Op. Cit.,* p. 359.
51. Alfred Price: *Luftwaffe* (New York: Ballantine, 1969), p. 87. For a detailed discussion of the issue, see "The Attack on the German Oil Industry: The Crippling of the Luftwaffe," in W. H. Tantum and E. J. Hoffschmidt (eds): *The Rise and Fall of the German Air Force* (Old Greenwich, Conn.: WE Inc., 1969), pp. 347–356.
52. E. E. Wigg: "Methanol as a Gasoline Extender: A Critique," *Science,* 29 November 1974, pp. 785–790. See also Thomas P. Hughes: "Technological Momentum in History: Hydrogenation in Germany 1898–1933," *Past and Present,* Vol. 44, August 1969, pp. 106–131.
53. David Irving: *Die Tragödie der Deutschen Luftwaffe* (Frankfort: Ullstein, 1970), p. 331.
54. *Ibid.,* p. 291.
55. Speer: *Op. Cit.,* p. 121.
56. *Ibid.,* p. 122.
57. David Irving: *The German Atomic Bomb* (New York: Simon and Schuster, 1967), p. 125
58. Werner Heisenberg: *Der Teil und das Ganze* (München: Piper, 1969), Chapter 6.
59. Ley: *Op. Cit.,* p. 245.
60. Frank Barnaby: "Thirty Years of Nuclear Weaponry," *New Scientist,* 7 August 1975, p. 330.
61. Bryan Cooper and John Batchelor: *Bombers: 1939—1945* (London: Phoebus, 1974), p. 33.
62. See Klee and Merk: *Op. Cit.,* p. 102 note 16, for a photo of this document.
63. Ruland: *Op. Cit.,* p. 169; Irving: *Mare's Nest,* p. 118.
64. *Ibid.,* p. 218.
65. John Cookson and Judith Nottingham: *A Survey of Chemical and Biological Warfare* (New York: Monthly Review Press, 1969), p. 220.
66. *Ibid.,* p. 219.
67. Ley: *Op. Cit.,* p. 271.
68. *The 1973 World Almanac* (New York: Newspaper Enterprise Association, 1972), p. 607.
69. Cookson and Nottingham: *Op. Cit.,* p. 173.

70. Irving: *Mare's Nest,* p. 169.
71. *Ibid.,* p. 287.
72. Cookson and Nottingham: *Op. Cit.,* p. 219.
73. Brian Ford: *German Secret Weapons* (New York: Ballantine, 1969), p. 111.
74. Lusar: *Op. Cit.,* p. 395.
75. Speer: *Op. Cit.,* p. 579.
76. *Ibid.,* p. 420.
77. Bernard Brodie and Fawn M. Brodie: *From Crossbow to H-Bomb* (Bloomington: Indiana University Press, 1973), p. 214.
78. Hitch and McKean: *Op. Cit.,* p. 224.
79. Speer: *Op. Cit.,* p. 579.
80. Simon: *Op. Cit.,* p. 191.
81. Winston S. Churchill: *The Hinge of Fate* (New York: Bantam, 1962), p. 99
82. Price: *Op. Cit.,* pp. 61–63.
83. Churchill: *Op. Cit.,* pp. 251–252.
84. *Ibid.,* p. 252.
85. *Ibid.,* p. 253.
86. John Killen: *A History of the Luftwaffe 1915-1945* (New York: Berkley, 1967) pp. 237–239.
87. Glenn H. Snyder: "The Balance of Power and the Balance of Terror," in Dean C. Pruitt and Richard C. Snyder: *Theory and Research on the Causes of War* (Englewood Cliffs, N.J.: Prentice-Hall, 1969), pp. 114–126.
88. Thomas C. Schelling: *The Strategy of Conflict* (London: Oxford University Press, 1960), p. 6.
89. *Ibid.,* p. 75.
90. Hitch and McKean: *Op. Cit.,* p. 202.
91. Speer: *Op. Cit.,* p. 421; Irving: *Tragödie,* p. 281.
92. Walter C. Langer: *The Mind of Adolf Hitler* (New York: Basic Books, 1972), p. 156.
93. For a useful discussion and valuable theory of such conversion experiences, see: Julian Silverman: "Shamans and Acute Schizophrenia," *American Anthropologist,* 1967, Vol. 69, pp. 21–31.
94. Irving: *Mare's Nest* p. 196.
95. *Ibid.,* p. 197.
96. *Ibid.*
97. Cookson and Nottingham: *Op. Cit.,* p. 312.
98. *Ibid.,* p. 296.
99. Karl-Heinz Ludwig: "Die Hochdruckpumpe—ein Beispiel technischer Fehleinschätzüng," *Technikgeschichte,* Vol. 38, No. 2, 1971, pp. 142–157.
100. Ley: *Op. Cit.,* p. 238.
101. Gerhard L. Weinberg (ed.): *Hitlers Zweites Buch–Ein Dokument aus dem Jahr 1928* (Stuttgart: Deutsche Verlags-Anstalt, 1961), p. 173.
102. Lusar: *Op. Cit.,* p. 403.
103. Ludwig: *Op. Cit.,* pp. 144–145.
104. Ludwig: *Op. Cit.,* p. 144.

105. Irving: *Mare's Nest,* pp. 121, 178, 213–219.
106. Ludwig: *Op. Cit.,* p. 146.
107. Irving: *Mare's Nest,* p. 215.
108. Ludwig: *Op. Cit.,* p. 401; Simon: *Op. Cit.,* p. 191.
109. Ludwig: *Op. Cit.,* p. 150.
110. Brian Ford: *German Secret Weapons* (New York: Ballantine, 1969), p. 142.
111. Irving: *Mare's Nest,* p. 246.
112. *Ibid.,* pp. 215, 218.
113. *Ibid.,* p. 248.
114. *Ibid.,* pp. 7, 8.
115. Lusar: *Op. Cit.,* pp. 209–210.
116. Ley: *Op. Cit.,* p. 605.
117. Dornberger: *V-2* (German edition), p. 267.
118. *Ibid.,* p. 268.
119. Ley: *Op. Cit.,* p. 266.
120. Lusar: *Op. Cit.* p. 209; Ford: *Op. Cit.,* p. 131, see also pp. 145–146 for a photograph and a blueprint of the rocket.
121. R. S. Hirsch and Uwe Feist: *Messerschmitt 262* (Fallbrook, Calif.: Aero, 1967), p. 5.
122. Killen: *Op. Cit.,* p. 268.
123. Eugene M. Emme: *Hitler's Blitzbomber* (Maxwell Air Force Base: Air University, 1951), p. 13; Price: *Op. Cit.,* p. 152.
124. Hirsch and Feist: *Op. Cit.,* p. 5.
125. Emme: *Op. Cit.,* p. 2.
126. John Killen: *A History of the Luftwaffe* (New York: Berkley, 1967) p. 271.
127. Ernst Heinkel: *Stürmisches Leben* (Stuttgart, Germany: Mundus, 1953).
128. Edward T. Maloney: *Heinkel He 162* (Fallbrook, Calif.: Aero, 1965), p. 5.
129. Killen: *Op. Cit.,* p. 271.
130. Speer: *Op. Cit.;* Heinkel: *Op. Cit.*
131. Killen: *Op. Cit.,* p. 271.
132. Ruland: *Op. Cit.,* p. 110.
133. Price: *Op. Cit.,* p. 128.
134. Emme: *Op. Cit.,* p. 20.
135. Killen: *Op. Cit.,* p. 272.
136. *Ibid.*
137. Hirsch and Feist: *Op. Cit.,* p. 6.
138. Speer: *Op. Cit.,* p. 416; Killen: *Op. Cit.,* p. 275.
139. Price: *Op. Cit.,* p.152.
140. Martin Caiden: *Me 109* (New York: Ballantine, 1968), p. 77.
141. Cooper and Batchelor: *Op. Cit.,* p. 55; Killen: *Op. Cit.,* pp. 288–289.
142. Lusar: *Op. Cit.,* pp. 160–161.
143. Ford: *Opl Cit., Op.* 95, 102.
144. Lusar: *Op. Cit.,* p. 162.
145. Hanna Reitsch: *Flying Is My Life* (New York: Putnam, 1954), pp. 207–219.

146. Ley: *Op. Cit.,* p. 600.
147. Klee and Merk: *Op. Cit.,* pp. 99, 100; Kenneth W. Gatland (ed.): *Project Satellite* (New York: British Book Centre, 1958), page facing 32.
148. Hans Hartl: *Hermann Oberth–Vorkämpfer der Weltraumfahrt* (Hannover, W. Germany: Oppermann, 1958), p. 192.
149. Ley: *Op. Cit.,* p. 272; von Braun and Ordway: *Op. Cit.,* p. 119.
150. Walter Dornberger: *V-2* (New York: Viking, 1954), p. 251.
151. Lehnert: "Die aerodynamische Entwicklung der Fernrakete A9 (A4b)," in an unpublished U.S. Army document, *The Story of Peenemünde,* 1945.
152. James C. Sparks: *Winged Rocketry* (New York: Dodd, Mead; 1968), p. 67.
153. Wernher von Braun: "From Small Beginnings...," in Gatland: *Op. Cit.,* 47–48.
154. Von Braun and Ordway: *Op. Cit.,* p. 119.
155. Cornelius Ryan (ed.): *Across the Space Frontier* (New York: Viking, 1952).
156. Richard S. Lewis: *Appointment on the Moon* (New York: Ballantine, 1969), p. 188.
157. Von Braun and Ordway: *Op. Cit.,* p. 145.
158. Speer: *Op. Cit.,* p. 375.
159. Dornberger: *V-2* (German edition), p. 278.
160. Kurzweg and Hermann: "Die aerodynamische Entwicklung der Flakrakete 'Wasserfall,'" (München: Wasserbau-Versuchsanstalt, 1945).
161. Klee and Merk: *Op. Cit.,* p. 65.
162. *Aviation Week,* 15 October 1973, pp. 12–18; Robert Gillette: "Military R & D: Hard Lessons of an Electronic War," *Science,* 9 November 1973, pp. 559–561.
163. *Aviation Week,* 8 Jan 1972, p. 19; 1 January 1972, p. 16.
164. Rowland F. Pocock: *German Guided Missiles* (New York: Arco, 1967), p. 82; see also p. 74.
165. Interview with Netzer in U.S. Army document, *The Story of Peenemünde.*
166. Pocock: *Op. Cit.,* p. 72.
167. Von Braun and Ordway: *Op. Cit.,* p. 145.
168. Ryan *Op. Cit.;* Wernher von Braun: *The Mars Project* (Urbana: University of Illinois Press, 1962).
169. "Lunar Orbiter," *NASA Facts,* Vol. IV, No. 4, 1967.
170. Raymond N. Watts: "Three Spacecraft Study the Red Planet," *Sky and Telescope,* January 1972, p. 15.
171. Simon: *Op. Cit.,* p. 126.
172. Lusar: *Op. Cit.,* p. 181.
173. Caiden: *Op. Cit.,* pp. 131–142.
174. Ley: *Op. Cit.,* p. 649.
175. *Ibid.,* p. 602.
176. Fritz Hahn: *Deutsche Geheimwaffen 1939–1945,* Volume I, "Flugzeugbewaffnungen" (Heidenheim, Germany: Hoffmann, 1963), p. 183.
177. *Ibid.,* pp. 180,183.
178. Simon: *Op. Cit.,* p. 52.
179. *Ibid.,* p. 49.
180. Hahn: *Op. Cit.,* p. 178.

181. Dornberger: *V-2* (German edition), p. 280.
182. Ford: *Op. Cit.,* p. 71.
183. Lusar: *Op. Cit.,* p. 208.
184. Killen: *Op. Cit.,* p. 289.
185. Ley: *Op. Cit.,* p. 518.
186. Sparks: *Op. Cit.,* p. 51.
187. Lusar: *Op. Cit.,* p. 209.
188. Ley: *Op. Cit.,* p. 519.
189. Alfred J. Zaehringer: *Soviet Space Technology* (New York: Harper, 1961), p. 76.
190. Horace Jacobs and Eunice Engelke Whitney: *Missile and Space Projects Guide* (New York: Plenum, 1962), p. 191.
191. *Ibid.,* p. 148.
192. Wernher von Braun: "The Redstone, Jupiter, and Juno," *Technology and Culture,* Fall 1963, pp. 452–465.
193. Nicholas Daniloff: *The Kremlin and the Cosmos* (New York: Knopf, 1972), p. 53.
194. For example, see Zaehringer: *Op. Cit.,* Plate X.
195. Evgeny Riabchikov: *Russians in Space* (Garden City, N.Y.: Doubleday, 1971), Plate 24.
196. J. Gordon Vaeth: *200 Miles Up* (New York: Ronald, 1956), p. 89.
197. *Ibid.,* p. 147.

Chapter 5

1. Sam Moskowitz: *Seekers of Tomorrow* (Westport, Conn.: Hyperion, 1966), p. 376; see also p. 136.
2. Harry Warner, Jr.: *All Our Yesterdays* (Chicago: Advent, 1969), p. 29.
3. Milton Lehman: *This Man High* (New York: Pyramid, 1970), pp. 149–150.
4. G. Edward Pendray: "Fifteen Years of Organized Rocketry," *Journal of the American Rocket Society,* June 1945, p. 9.
5. G. Edward Pendray: "The First Quarter Century of the American Rocket Society," *Jet Propulsion,* November 1955, p. 587.
6. Donald B. Day: *Index to the Science-Fiction Magazines, 1926–1950* (Portland, Ore.: Perri Press, 1952).
7. *Ibid.,* p. 38.
8. *Ibid.,* pp. 45, 63.
9. *Ibid.,* pp. 50–51.
10. *Ibid.,* p. 45
11. *Ibid.,* pp. 58–59.
12. *Ibid.,* p. 39.
13. *Ibid.,* p. 20.
14. Sam Moskowitz: *Seekers of Tomorrow,* p. 104; *The Immortal Storm* (Westport, Conn.: Hyperion, 1974), p. 11.
15. *Bulletin of the American Interplanetary Society,* June 1930, p. 1.

16. "The A. R. S. Journal," *Journal of the American Rocket Society,* No. 61, March 1945, p. 14.
17. Moskowitz: *The Immortal Storm,* p. 18.
18. Donald H. Tuck: *The Encyclopedia of Science Fiction and Fantasy* (Chicago: Advent, 1974), Volume 1, p. 265.
19. *Bulletin of the AIS,* February 1931, p. 1.
20. P. E. Cleator: *Rockets through Space* (New York: Simon and Schuster, 1936), p. 134.
21. *Bulletin of the AIS,* May 1931, p. 5.
22. Pendray: "Fifteen Years of Organized Rocketry," p. 10.
23. Lehman: *Op. Cit.,* p. 151.
24. *Journal of the British Interplanetary Society,* October 1935, p. 9.
25. *Ibid.,* p. 11.
26. G. Edward Pendray: "The Conquest of Space by Rocket," *Bulletin of the AIS,* March 1932, p. 7.
27. Erik Bergaust and William Beller: *Satellite!* (Garden City, N.Y.: Hanover House, 1956), p. 36; *Astronautics,* March 1934, p. 7.
28. Pendray: "The First Quarter Century of the American Rocket Society," p. 590.
29. See: A. V. Cleaver: "Nose Drive or Tail Drive?" *Journal of the BIS,* May 1949, pp. 120–123.
30. Andrew G. Haley: *Rocketry and Space Exploration,* (Princeton, N.J.: Van Nostrand, 1958), pp. 21–22.
31. *Journal of the BIS,* Oct. 1935, p. 10.
32. G. Edward Pendray: "The History of the First A. I. S. Rocket," *Astronautics,* November–December 1932, p. 2.
33. G. Edward Pendray: "The Flight of Experimental Rocket No. 2," *Astronautics,* May 1933, pp. 10–11.
34. *Ibid.,* p. 13.
35. J. H. Wyld: "Fuel as Coolant," *Astronautics,* April 1938, pp. 11–12.
36. John Shesta, H. Franklin Pierce, and James H. Wyld: "Report on the 1938 Rocket Motor Tests," *Astronautics,* February 1939, p. 6.
37. *Astronautics,* February 1939, pp. 11–16.
38. *Astronautics,* August 1941, pp. 3–4.
39. G. Edward Pendray: "32 Years of ARS History," *Astronautics and Aerospace Engineering,* February 1963, p. 126.
40. Von Braun and Ordway: *Op. Cit.,* p. 94.
41. Pendray: *The Coming Age of Rocket Power,* p. 42.
42. Bergaust and Beller: *Op. Cit.,* p. 36.
43. Lovell Lawrence: "The Reaction Motors Model 6000C4 Rocket Engine," *Journal of the BIS,* December 1947, p. 197.
44. "Index to Articles in 'Astronautics,' " *Journal of the American Rocket Society,* No. 68, December 1946, pp. 41–46.
45. "The A. R. S. Journal," *Journal of the American Rocket Society,* No. 61, March 1945, p. 14.
46. *Astronautics,* March 1944, p. 15.
47. Roy Healy writing in *Astronautics,* February 1943, p. 10.

48. *Astronautics,* December 1944, p. 14.
49. G. Edward Pendray: "Society Affiliates with A. S. M. E.," *Journal of the American Rocket Society,* No. 64, December 1945, p. 30.
50. "Roster of Active Members, 15 October 1946," *Journal of the American Rocket Society,* No. 68, December 1946, pp. 17–22.
51. *Journal of the American Rocket Society,* March–April 1953, p. 103.
52. *Journal of the American Rocket Society,* July–August 1952, p. 191.
53. See editorial, *Journal of the American Rocket Society,* November–December, 1953, p. 337.
54. Martin Summerfield: "We Change Our Name," *ARS Journal,* January 1959, p. 7.
55. Bergaust and Beller: *Op. Cit.,* pp. 36–37.
56. H. K. Wilgus: "Final Report of the Ad Hoc Space Flight Committee of the American Rocket Society," *Journal of the American Rocket Society,* March–April 1953, pp. 103–104.
57. *Jet Propulsion,* February 1955, pp. 71–78.
58. Ley: *Op. Cit.,* p. 588.
59. *Astronautics,* Journal of the American Astronautical Society, Fall 1954, Vol. I, No. I, p. 20.
60. *Ibid.*
61. *The Journal of Astronautics,* Summer 1955, p. 73.
62. Jerry Grey and Vivian Grey: *Space Flight Report to the Nation* (New York: Basic Books, 1962).
63. *Journal of the American Rocket Society,* March–April 1952, p. 104.
64. C. J. McCarthy: "IAS: A Capsule History," *Astronautics and Aerospace Engineering,* February 1963, p. 127.
65. Robert Hotz: "Filling the Vacuum" *Aviation Week,* 24 September 1973, p. 7.
66. John Tormey: "A Member's Report Card on AIAA's Publications," *Astronautics and Aeronautics,* January 1969, pp. 58–63.
67. *AIAA Roster 1973* (New York: AIAA, 1973).
68. *Astronautics and Aeronautics,* November 1973, p. 25.
69. Gerd Hortleder: *Das Gesellschaftsbild des Ingenieurs* (Frankfurt: Suhrkamp, 1970).
70. Ley: *Op. Cit.,* p. 144.
71. H. Lorenz: "Die Möglichkeit der Weltraumfahrt," *Zeitschrift des Vereines deutscher Ingenieure,* Vol. 71, No. 19, May 1927, pp. 651–654; continued p. 1128.
72. *Ibid.,* p. 653; percentages calculated from Table 3.
73. *Ibid.,* p. 1236.
74. Ley: *Op. Cit.,* pp. 144, 145.
75. "This Is AIAA," xerox of typewritten copy provided by Wheeler X. Johnson, Treasurer of the New England Section of the AIAA, 1972.
76. *Astronautics and Aeronautics,* November 1973, p. 26.
77. John H. Sidebottom: "Finances," *Astronautics and Aeronautics,* March 1974, pp. 29–32.
78. Charles V. Willie: Form letter on behalf of the Eastern Sociological Society, Fall 1974.
79. *Astronautics and Aeronautics,* November 1973, p. 29.

80. *Ibid.,* p. 25.
81. See Article VI, sections 6.1 through 6.4 of the AIAA Constitution, revised January 10, 1973.
82. *Publications of the AIAA: 1974 Subject and Author Indexes* (New York: AIAA, 1975).
83. Ley:*Op. Cit.,* p. 94.
84. Frank H. Winter: "Sir William Congreve: A Bicentennial Memorial," *Spaceflight,* September 1972, pp. 333–334.
85. *Ibid.*
86. Ley: *Op. Cit.,* p. 93.
87. Frank H. Winter: "William Hale—A Forgotten British Rocket Pioneer," *Spaceflight,* January 1973, pp. 31–33.
88. Ley: *Op. Cit.,* pp. 96–97.
89. Winter, *Op. Cit.*
90. *Ibid.,* p. 32.
91. Ley: *Op. Cit.,* p. 589.
92. Von Braun and Ordway: *Op. Cit.,* p. 76.
93. P. E. Cleator: "Autopsia," in Arthur C. Clarke: *The Coming of the Space Age* (New York: Meredith, 1967), pp. 67–70.
94. P. J. Parker: Letter, *Spaceflight,* July 1972, pp. 277–278.
95. Von Braun and Ordway: *Op. Cit.,* p. 77.
96. Cleator: *Rockets through Space,* pp. 136–139.
97. Moskowitz: *Seekers of Tomorrow,* p. 137.
98. *Day's Index,* p. 13.
99. *Journal of the BIS,* January 1934, p. 3.
100. *Journal of the BIS,* September 1951, p. 234.
101. *Journal of the BIS,* January 1934, p. 3.
102. P. E. Cleator: "Autopsia," *Journal of the BIS,* May 1948, p. 97.
103. A. V. Cleaver: "The Post-War Contribution of the B. I. S." *Spaceflight,* September 1961, p. 169.
104. H. E. Ross: "Gone with the Efflux," *Journal of the BIS,* May 1950, p. 101.
105. P. E. Cleator: "Matters of no Moment," *Journal of the BIS,* March 1950, p. 51.
106. *Journal of the BIS,* Oct. 1934, p. 34.
107. *Journal of the BIS,* May 1935, p. 2.
108. Cleator: *Rockets through Space,* cited in note 20.
109. *Journal of the BIS,* June 1936, p. 36.
110. *Journal of the BIS,* February 1937, p. 4.
111. *Things to Come,* also titled *The Shape of Things to Come,* Great Britain, 1936. Directed by William Cameron Menzies, produced by Alexander Korda, screenplay by H. G. Wells. John Baxter: *Science Fiction in the Cinema* (New York: Paperback Library, 1970), p. 61. My own transcription of the film's sound track does not agree exactly with Baxter's version, but the differences are not important.
112. Arthur C. Clarke (quoting a prewar leaflet he himself had written): "The Challenge of the Spaceship," *Journal of the BIS,* December 1946, p. 67.
113. *Time,* 25 July 1969, p. 17.

114. "Astronautics in Britain," *Spaceflight,* June 1967, p. 201.
115. Arthur C. Clarke: *Voices from the Sky* (New York: Harper & Row, 1965), p. 168.
116. *Ibid.*
117. *Journal of the BIS,* December 1937, p. 5.
118. *Journal of the BIS,* January 1939, July 1939.
119. See the improved reprint of one of the original articles: H. E. Ross: "The B. I. S. Spaceship," *Spaceflight,* February 1969, p. 42; and: Dave Dooling: "The Evolution of the Apollo Spacecraft," *Spaceflight,* March 1974, p. 82.
120. R. A. Smith: "The B. I. S. Coelostat," *Journal of the BIS,* July 1939, pp. 22–27.
121. See autobiographical sketches by Clarke and Cleaver in Shirley Thomas: *Men of Space* (New York: Chilton, 1968), Volume 8, pp. 14–20.
122. "The British Interplanetary Society," membership booklet, p. 5.
123. Arthur C. Clarke: *Voices from the Sky,* p. 175.
124. C. S. Lewis: *Perelandra* (New York: Macmillan, 1965), pp. 81–82.
125. Patrick J. Callahan: "The Two Gardens in C. S. Lewis's *That Hideous Strength,*" in Thomas D. Clareson (ed.): *SF: The Other Side of Realism* (Bowling Green, Ohio: Bowling Green University Popular Press, 1971), p. 155.
126. C. S. Lewis: *That Hideous Strength* (New York: Macmillan, 1965), pp. 178–179.
127. C. S. Lewis: *Out of the Silent Planet* (New York: Macmillan, 1965), pp. 32, 154.
128. Ley: *Op. Cit.,* p. 589.
129. Shirley Thomas: *Men of Space,* Volume 8, p. 3.
130. *BIS Annual Report–1947.*
131. P. E. Cleator: "Autopsia," p. 98.
132. Thomas: *Op. Cit.,* Volume 8, pp. 18–19.
133. Michael N. Golovine: "Space—The New Spearhead," *Spaceflight,* November 1965, p. 181.
134. *Journal of the BIS,* July 1951, p. 145.
135. C. E. Tharratt: "Personal Profile," *Spaceflight,* August 1972, p. 305.
136. A. V. Cleaver: "The Scope and Extent of UK Participation in Space," *Spaceflight,* March 1972, pp. 99–102.
137. "Major U. K. Rocketry Ends with Europa II," *Spaceflight,* September 1973, pp. 332–333.
138. A. V. Cleaver: "The Scope and Extent," p. 99; note 135.
139. Donald E. Fink: "Europeans See Wide Use of Spacelab," *Aviation Week,* 5 November 1973, pp. 42–46. See also: William A. Schumann: "NASA Shuttle Projection Sees Heavy Spacelab Use," *Aviation Week,* 5 November 1973, p. 22; "NASA and ESRO: A European Payload for the Space Shuttle," *Science,* 9 November 1973, p. 562; Richard Lewis: "ESRO's Cabin in the Sky," *New Scientist,* 27 September 1973, pp. 767–768; Kenneth Gatland: "Missing the Space Boat," *New Scientist,* 12 October 1972, p. 68.
140. A. V. Cleaver: "The Scope and Extent," p. 99; note 135.
141. "The British Interplanetary Society," membership booklet, p. 5.
142. *Ibid.,* p. 2.
143. "The Report of the Council," *Spaceflight,* May 1973, pp. 196–200.
144. A "Space Study Meeting" with the theme "Project Starship" was held in London by

the BIS in January of 1973, according to an advertişement on the back cover of the December 1972 issue of *Spaceflight.*

145. Kenneth Gatland: "Trekking to Barnard's Star," *New Scientist,* 29 August 1974, pp. 522–523; "Project Daedalus," *Spaceflight,* September 1974, pp. 356–358.

146. *Spaceflight,* January 1974, p. 1.

Chapter 6

1. Barbara Hubbard: "From Meaninglessness to New Worlds," *The Futurist,* April 1971, p. 72.
2. *Ibid.*
3. Barbara Hubbard: "The Shift from Revolutionary to Evolutionary Activism," *ARISE,* Committee for the Future, Philadelphia, 1972.
4. Abraham H. Maslow: *Religions, Values and Peak-Experiences* (New York: Viking, 1970).
5. Barbara Hubbard: "From Meaninglessness," p. 73.
6. *Ibid.*
7. Pierre Teilhard de Chardin: *The Future of Man* (New York: Harper & Row, 1964
8. Earl Hubbard: *The Search Is On* (Los Angeles: Pace Publications, 1969), p. 156. Since Pace Publications has gone out of business, this book is only available from The World Future Society which has a book service. Earl Hubbard's writings have never received extensive distribution.
9. *Ibid.,* p. 22.
10. *Ibid.,* p. 126.
11. *Ibid.,* p. 122.
12. *Ibid.,* p. 85.
13. Earl Hubbard: "Who Are 'We'?—A Theotropic Form of Energy," *New Worlds,* December 1971, pp. 16–26.
14. Earl Hubbard: *The Search Is On,* p. 120.
15. Henri Bergson: *Creative Evolution* (New York: Holt, 1911).
16. Earl Hubbard: *The Search Is On,* p. 98.
17. Earl Hubbard: "The Need for New Worlds," *New Worlds,* April 1971, p. 13.
18. Lecture by Dr. Roger W. Wescott at the May 1973 Syncon in Washington, D.C., tape recorded.
19. Tape recording of the performance of *The Pyramid* given at the May 1972 Syncon at Southern Illinois University, Carbondale; tape recording of *Marut: The First Thousand Years,* given at the May 1973 Syncon in Washington, D.C.
20. For a late and summarizing example of occult pyramid-measuring, see H. Spencer Lewis: *The Symbolic Prophecy of the Great Pyramid* (San Jose, Calif.: Supreme Grand Lodge of AMORC, 1963). For a good critique of this occult activity, see: Martin Gardner: *Fads and Fallacies in the Name of Science* (New York: Dover, 1957), pp. 173–185.
21. Hermann Oberth: *Stoff und Leben* (Remagen, Germany: Reichl, 1959).
22. Earl Hubbard: "Mankind in the Universe," *The Futurist,* Vol. 1, No. 6, December 1967, pp. 90–92.

23. See photograph, *The Futurist,* Vol. 2, No. 1, February 1968, p. 13.
24. See list of staff, *The Futurist,* Vol. 2, No. 1, February 1968, p. 2. The articles published by the Hubbards after this time in the magazine of The World Future Society were Barbara Hubbard: "Humanistic Psychology and the 'New Questions,' " *The Futurist,* April 1968, pp. 32–33; "Lancelot Law Whyte and the 'Integrators,' " *The Futurist,* August 1968, p. 80; "Pollyannaism in Reverse: A Threat to Man's Future," *The Futurist,* December 1968, p. 131; "From Meaninglessness to New Worlds," *The Futurist,* April 1971, pp. 72–75; "An Appetite for Evolution," *The Futurist,* February 1973, pp 26–29. Earl Hubbard: "The Need for New Worlds," *The Futurist,* October 1969, pp. 117 –118.
25. Richard A. Givens: "The Mass Media and the Future," *The Futurist,* December 1968, p. 131.
26. James C. Sparks: *Winged Rocketry* (New York: Dodd, Mead, 1968).
27. Interview with Sister Mary Fidelia, May 1973; Henry Libersat: "Science, Faith Seen as Mutual Benefit," *New Worlds,* February 1972, pp. 12–13.
28. *New Worlds,* April 1971, p. 2.
29. *Ibid.*
30. "A Case for Harvest Moon," *New Worlds,* April/May 1972, p. 23; CFF Fact Sheet 71-722-203.
31. This is ridiculous. The lunar rover's batteries are surely dead, the scientific package must be left undisturbed, and the descent stage will be found empty and perhaps damaged by the ascent stage take-off.
32. CFF Fact Sheet 72-215-105. The special sources I used in preparing this section on Harvest Moon were:
    CFF Fact Sheet 71-722-203; CFF Fact Sheet 72-66-107; CFF Fact Sheet 72-215-105; undated CFF bulletin, approximately August 1972; untitled CFF xerox announcement, May 20, 1972; New Worlds Newsletter, Vol. 2, No. 1, January 1974, p. 4; New Worlds Newsletter, undated, approximately August 1972; New Worlds Newsletter, undated, approximately December 1972; "Status Report, Project: Harvest Moon," *New Worlds,* December 1971, pp. 13–15; "A Case for Harvest Moon," *New Worlds,* April/May 1972, pp. 21–24; "Harvest Moon Goes Global," *New Worlds,* January 1972, p. 6.
    And the public sources:
    1972 NASA Authorization, GPO, Washington, 1972, pp. 648–650; "House Resolution 979," May 11, 1972.
33. Gene Simmons: *On the Moon with Apollo 16* (Washington, D.C.: NASA, 1972), pp. 39–40.
34. An important second step was to be the founding of the first true lunar colony by 1982.
35. See "76 International Skylab?" *Astronautics & Aeronautics,* September 1972, p. 5.
36. A representative of North American Rockwell gave a small, informal presentation on the possibility of this project at the May 1972 Syncon, but the Committee paid no attention. See "Cooperation in the Cosmos," *Time,* 5 June 1972, p. 19.
37. Turner has told me that the breakup of the Fuller organization, already proceeding apace at the time, was caused by an Illinois political struggle, rather than internal problems. S. I. U. faculty member John Gardner has written an interesting report on the university, of which he said, "Experimentation, daring innovation have been S. I. U.'s trademark"; John Gardner: "We Teach and Study and Raise all the Hell

We Can," *Change,* June 1973, pp. 43–48. Since this time, S. I. U. has even experimented with mass firing of tenured faculty!

38. Barbara Hubbard: "From Meaninglessness to New Worlds," *The Futurist,* April 1971, p. 72.
39. Donella Meadows et al.: *The Limits to Growth* (New York: Basic Books, 1972).
40. Krafft Ehricke: "Extraterrestrial Imperative—Road into the Future," North American Rockwell document SD72-SA-0120, p. 30.
41. Shirley Thomas: *Men of Space* (Philadelphia: Chilton, 1960), Volume I, p. 4.
42. N. A. Rynin: *Interplanetary Flight and Communication,* Vol. I, No. 3 (Leningrad 1931), Israel Program for Scientific Translations, Jerusalem, 1971, pp. 112–132.
43. Presentations on the Limits to Growth issue continued to be made, but in the context of the Syncon process, not as an independent project.
44. *New Worlds Newsletter,* January 1974, p. 4.
45. Because this Syncon involved a variety of satellite groups, the wall removal did not proceed in the most simple fashion, and some groups moved from one section to another to be with the right partner task forces. It seems the pattern at later Syncons has also been unbalanced. At the May 1972 Carbondale Syncon, the six-section Wheel had permitted a simpler amalgamation process. At that meeting when the first Walls Down time came, every other internal partition of the Wheel was removed, reducing the number of sections from six to three. In Washington a lopsided four-section Wheel was the intermediary result.
46. The Southern Illinois University newspaper, *Daily Egyptian,* Carbondale, May 19, 1972, pp. 1 and 15.
47. Howard Muson: "Comedown from the Moon—What Has Happened to the Astronauts," *New York Times Magazine,* 3 December 1972, p. 134.
48. *Boston Sunday Globe,* 16 July 1972, p. 2.
49. *Boston Sunday Globe,* 1 July 1973, p. A-2.
50. See note 47, p. 37.
51. Paul Langner: "Meditation Called Savior of Urban Man," *Boston Sunday Globe,* 6 July 1972, p. 28.
52. Robert Weiss: "New Worlds Video," *New Worlds,* April/May 1972, pp. 15–17; *New Worlds,* February 1972, p. 4.
53. When I inventoried the equipment in use at the May 1973 Syncon, I found the van and monitors, but only three color cameras and 6 black and white cameras.
54. CFF booklet, "Syncon—Washington, D.C.," p. 4.
55. A brief summary of a number of studies is given in David Krech, Richard S. Crutchfield, and Egerton Ballachey: *Individual in Society* (New York: McGraw-Hill, 1962), pp. 295–261. An entire issue of *The Journal of Nervous and Mental Disease* has been devoted to the use of television and video tape in personality or behavior change programs. See for example, Ian Alger: "Therapeutic Use of Videotape Playback," Vol. 148, No. 4, pp. 430–436.
56. A handout said Syncomp would allow participants to "join or start a discussion/read or make proposals/get a list of participants with similar interests/comment on SYNCON/announce your own ad-hoc session/send a message. SYNCOMP adds a new dimension to SYNCON by helping you to rapidly discover those people with common interests and set up a continuing dialogue over several days with large numbers of people on a wide variety of topics."

57. For example, here is part of the Information Evolution output: "Intro: Two Main Problems: "Information Pollution" "Meta-information crisis—where to get info" goals: Solve above Problems. Needs: (1) Establish and define socially acceptable limits of privacy. Resources: All necessary resources available. (2) Meta-information crisis—We need information about information—information may be available about "X" but one doesn't know where to go to get that information about "X." Goals: Solve problems 1 and 2...."
58. Michael Kernan: "Honorable Intentions," *The Washington Post,* Friday, 11 May 1973, pp. B-1 and B-10.
59. Georg Simmel: *The Sociology of Georg Simmel* (New York: Macmillan, 1950), pp. 34–39, 227–229.
60. The questionnaire was published in *New Scientist,* 23 November 1972, p. 465, and the results in *New Scientist,* 25 January 1973, p. 209.
61. Mark Wynn: "Who Are the Futurists?" *The Futurist,* April 1972.
62. "Giving Mankind a Future," *New Worlds,* december 1971, p. 15.
63. CFF mimeographed handout: "Community for the Future," May 1973.
64. Peter L. Berger: *Invitation to Sociology* (Garden City, N.Y.: Doubleday, 1963), p. 147.
65. Peter L. Berger and Thomas Luckmann: *The Social Construction of Reality* (Garden City, N.Y.: Doubleday, 1967), p. 102.
66. Peter L. Berger: *The Sacred Canopy* (Garden City, N.Y.: Doubleday, 1969), p. 21.
67. Viktor E. Frankl: *Man's Search for Meaning* (New York: Washington Square, 1963), p. 59.
68. Pitirim A. Sorokin: *The Ways and Power of Love* (Chicago: Regnery, 1963), p. 11.
69. Emile Durkheim: *Suicide* (New York: Free Press, 1951 [1897]), p. 248.
70. Subtitle of a Huntsville tourist brochure, "Rockets to the Stars" (Florence, Ala. Curt Teich & Co., Chicago/Anderson News Co., 1962).

Chapter 7

1. Sam Moskowitz: *Explorers of the Infinite* (New York: World, 1963), p. 225.
2. Sam Moskowitz: *The Immortal Storm* (Westport, Conn.: Hyperion, 1974), pp. 4 and 5.
3. Arthur C. Clarke: *Profiles of the Future* (New York: Bantam Books, 1964), p. xii. This book is dedicated to Hugo Gernsback.
4. Damon Knight: *In Search of Wonder* (Chicago: Advent, 1967), p. 90.
5. Alva Rogers: *A Requiem for Astounding* (Chicago: Advent, 1964).
6. N. W. Ayer and Sons: *American Newspaper Annual and Directory,* Philadelphia; this annual has appeared under different titles.
7. Donald B. Day: *Index to the Science Fiction Magazines, 1926–1950* (Portland, Ore.: Perri Press, 1952); Erwin S. Strauss: *Index to the SF Magazines, 1951–1965* (Cambridge, Mass.: MIT Science Fiction Society, 1966); *Index to the Science Fiction Magazines, 1966–1970* (Cambridge, Mass.: New England Science Fiction Association, 1971); *The N.E.S.F.A. Index, Science Fiction Magazines and Original Anthologies 1971–1972* (Cambridge, Mass., 1973); *The N.E.S.F.A. Index to the Science Fiction Magazines: 1973* (Cambridge, Mass., 1974). The most recent volumes are not so well

prepared as the earlier ones. The nature of many of the entries cannot be determined; and often even the correct date is missing.

8. Rogers: *Op. Cit.*, p. 12.
9. Moskowitz: *The Immortal Storm*, pp. 194–195.
10. *Ibid.*
11. Theodore Peterson: *Magazines in the Twentieth Century* (Urbana: University of Illinois Press, 1956), p. 286.
12. Rogers: *Op. Cit.*, p. 213.
13. Sam Moskowitz: *Seekers of Tomorrow* (Westport, Conn.: Hyperion, 1974), p. 424.
14. Winthrop Sargent: "Through the Interstellar Looking Glass," *Life*, 21 May 1951, p. 130.
15. *Ibid.*, p. 127.
16. "Science Fiction Rockets into Big Time," *Business Week*, 20 October 1951, p. 89.
17. "Progress in Science-Fiction: No Boom, But a Solid Market," *Publishers' Weekly*, 11 August 1951, p. 546.
18. Judith Merril: "What Do You Mean: Science Fiction?" in Thomas D. Clareson: *SF: The Other Side of Realism* (Bowling Green, Ohio: Bowling Green University Press, 1971), p. 74.
19. John Baxter: *Science Fiction in the Cinema* (New York: Paperback Library, 1970), pp. 215, 101 and 102.
20. Leonard Maltin (ed.): *TV Movies* (New York: Signet, 1969).
21. *The 1973 World Almanac*, p. 1032.
22. Ron Goulart: *An Informal History of the Pulp Magazines* (New York: Ace, 1972).
23. Robert M. Overstreet: *The Comic Book Price Guide–1974* (Cleveland, Tenn.: Overstreet, 1974).
24. These data are published every year in the convention program booklets; I took the figures from the Noreascon and Lacon booklets, 1971 and 1972.
25. Harry Warner, Jr.: *All Our Yesterdays* (Chicago: Advent, 1969), p. 25.
26. Robert A. W. Lowndes: "The Influence of Fandom," *Algol*, No. 17, 1971, p. 4.
27. *ApaNesfa*, No. 16.
28. *Proper Boskonian*, No. 3, p. 6.
29. *Luna Monthly*, 26/27, p. 38.
30. John W. Campbell, J.: "Letter from the Editor" in Rogers: *Op. Cit.*, p. xix.
31. Warner: *Op. Cit.*, p. 39.
32. Scope of fanzine sample:

Fanzine	*Issues*	*Pages*
Proper Boskonian	7	345
Luna Monthly	7	276
Algol	2	88
Granfalloon	6	296
Instant Message	24	156
Locus	40	424
ApaNesfa	1	36
Total		1,621

33. Donald Gould: "Life in Vitro," *New Scientist*, 13 January 1972, p. 100.
34. Robert A. Heinlein: "Science Fiction: Its Nature, Faults and Virtues," in Basil

Davenport (ed.): *The Science Fiction Novel* (Chicago: Advent, 1969), p. 26.

35. *Ibid.*, p. 24.
36. Alexei Panshin: *Heinlein in Dimension* (Chicago: Advent, 1968), p. 24.
37. Rogers: *Op. Cit.*, p. 133.
38. Arthur C. Clarke: *Voices from the Sky* (New York: Harper & Row, 1965), pp. 119–128.
39. *Ibid.*, pp. 233–241.
40. *Ibid.*, p. 125.
41. James Dugan: *Man under the Sea* (New York: Harper, 1956), p. 87.
42. *Ibid.*, p. 28.
43. Arthur C. Clarke: *The Exploration of Space* (Greenwich, Conn.: Fawcett, 1959), p. 15.
44. Larry Niven: *Ringworld* (New York: Ballantine, 1970), pp. 77, 81.
45. Martin Gardner: *Fads and Fallacies in the Name of Science* (New York: Dover, 1957), p. 19.
46. *Ibid.*, p. 21. See also: Willy Ley: "The Hollow Earth," *Galaxy*, March 1956, pp. 71–81.
47. Moskowitz: *Explorers of the Infinite*, p. 54.
48. *Ibid.*, p. 77.
49. George H. Waltz, Jr.: *Jules Verne–The Biography of an Imagination* (New York: Holt, 1943), p. 186.
50. I. O. Evans: *Jules Verne and His Work* (London: Arco, 1965), p. 45.
51. Richard A. Lupoff: *Edgar Rice Burroughs: Master of Adventure* (New York, Ace, 1965), pp. 66–70.
52. Freeman J. Dyson: "Search for Artificial Stellar Sources of Infrared Radiation," in A. G. W. Cameron: *Interstellar Communication* (New York: Benjamin, 1963), pp. 111–114.
53. *Ibid.*, p. 114.
54. N. A. Rynin: *Interplanetary Flight and Communication*, No. 7 (Jerusalem: Israel Program for Scientific Translations, 1971), pp. 24–25.
55. Richard S. Lewis: *Appointment on the Moon* (New York: Ballantine, 1969), p. 368.
56. Ryan (ed.): *Op. Cit.*, pp. 106–107 and the endpapers give excellent views of the space station, clearly a development of Noordung's idea, as is admitted on p. 99. The criticisms of Noordung made by Ley on that page are excessive. The current trend in space station design, as expressed in Salyut and Skylab, does not call for a large rotating wheel.
57. Hermann Noordung: *Das Problem der Befahrung des Weltraums* (Berlin: Schmidt, 1928).
58. For example, the card catalog of Harvard's Houghton rare book library.
59. Day's *Index*, p. 48.
60. See Day's *Index*.
61. Jonathan Norton Leonard: *Flight into Space* (New York: Random House, 1957), p. 284.
62. *Gallup Opinion Index*, March 1969, p. 17.
63. Irene Taviss: "A Survey of Popular Attitudes toward Technology," *Technology and Culture*, October 1972, p. 617.
64. *Proper Boskonian*, February 1968.
65. Stephen E. Whitfield: *The Making of Star Trek* (New York: Ballantine, 1968), p. 395.
66. David Gerrold: *The World of Star Trek* (New York: Ballantine, 1973), p. 157.

67. *Analog,* December 1971, p. 52.
68. Moskowitz: *The Immortal Storm,* p. 39. ff.
69. *Ibid.,* pp. 119, 213–224.
70. Robert A. Lowndes: "A Eulogy for the Dying Science-Fiction Magazines," *Riverside Quarterly,* Vol. 6, No. 1, August 1973, p. 34.
71. Damon Knight: *In Search of Wonder* (Chicago: Advent, 1967), p. 90.
72. Brian W. Adliss: *Billion Year Spree* (Garden City, N.Y.: Doubleday, 1973), p. 245.
73. Warner: *Op. Cit.,* pp. 186, 187.
74. *Ibid.,* pp. 32, 186.
75. *Ibid.,* p. 32.
76. William Bainbridge: *The Scientology Game,* unpublished undergraduate honors thesis, Boston University, 1971.
77. L. Ron Hubbard: *Dianetics–The Modern Science of Mental Health* (New York: Paperback Library, 1950), p. 9.
78. Warner: *Op. Cit.,* pp. 75–78.
79. Aldiss: *Op. Cit.,* p. 221.
80. Moskowitz: *Explorers of the Infinite,* p. 11.
81. William Sims Bainbridge: *Deviant Knowledge,* unpublished theory paper, Harvard University, 1973.
82. C. E. M. Hansel: *ESP—A Scientific Evaluation* (New York: Scribner's, 1966).
83. Alva Rogers: *A Requiem for Astounding,* pp. 20, 186.

Chapter 8

1. William Irwin Thompson: *At the Edge of History* (New York: Harper, 1971), p. ix.
2. Macfarlane Burnet: "After the Age of Discovery,"*New Scientist,* 9 December 1971, pp. 96–100. Alan Musset: "Discovery: A Declining Asset?" *New Scientist,* 27 December 1973, pp. 886–889.
3. Adrian Berry: *The Next Ten Thousand Years* (New York: Mentor, 1974) contains a number of technical errors. Dandridge M. Cole: *Beyond Tomorrow* (Amherst, Wisc.: Amherst Press, 1965).
4. Duncan Lunan: *Interstellar Contact* (Chicago: Regnery, 1974).
5. Craig, Covault: "Outer Planet Missions Keyed to Centaur," *Aviation Week,* 9 June 1975, p. 48.
6. *New York Times,* 7 January 1973, p. 1.
7. National Science Foundation news release NSF 72-199, mailed December 28, 1972.
8. *Aviation Week,* January 29, 1973.
9. Dale P. Cruikshank and David Morrison: "Titan and Its Atmosphere," *Sky and Telescope,* August 1972, pp. 83–85. Garry Hunt: "The Riddle of Titan's Gas," *New Scientist,* 7 November 1974, pp. 429, 430. A. G. W. Cameron: "The Outer Solar System," *Science,* 18 May 1973, pp. 701–708.
10. Robert Salkeld: "Single-Stage Shuttles for Ground Launch and Air Launch," *Astronautics and Aeronautics,* March 1974, pp. 52–64.
11. For a particularly frightening statement see David Dinsmore Comey: "The Legacy of Uranium Tailings," *Bulletin of the Atomic Scientists,* September 1975, pp. 43–45.

12. Derek P. Gregory: "Hydrogen—Transportable Storage Energy Medium," *Astronautics and Aeronautics,* August 1973, pp. 38–43.
13. G. Daniel Brewer: "The Case for Hydrogen-Fueled Transport Aircraft," *Astronautics and Aeronautics,* May 1974, pp. 40–51. Erwin J. Bulban: "Liquid Hydrogen Studied for Transports," *Aviation Week,* 5 November 1973, pp. 27–28. Craig Covault: "Fuel Shortages Spur Hydrogen Interest," *Aviation Week,* 17 December 1973, pp. 38–42.
14. William J. D. Escher: "Future Availability of Liquid Hydrogen," *Astronautics and Aeronautics,* May 1974, pp. 55–59.
15. Sebastian von Hoerner: "The General Limits of Space Travel," in A. G. W. Cameron: *Interstellar Communicaton* (New York: Benjamin, 1963), p. 156.
16. Eugen Sänger: *Raumfahrt–Technische Überwindung des Krieges* (Hamburg, Germany: Rowohlt, 1958), p. 90.
17. Philip J. Klass: "Major Hurdles for Laser Weapons Cited," *Aviation Week,* 9 July 1973, pp. 38–41. See also C. Frederick Hansen and George Lee: "Laser Power Stations in Orbit," *Astronautics and Aeronautics,* July 1972, pp. 42–55.
18. *NASA Activities,* July 1975, p. 5.
19. The reader should consult the most recent introductory Astronomy text. The September 1975 issue of *Scientific American* is devoted to the solar system and is the most recent overview of the subject at the time of this writing.
20. Roger A. MacGowan and Frederick I. Ordway: *Intelligence in the Universe* (Englewood Cliffs, N.J.: Prentice-Hall, 1966), especially p. 235.
21. Shirley Thomas: *Men of Space,* (Philadelphia: Chilton, 1960), Volume I, p. 21.
22. H. Spencer Jones: *Life on Other Worlds* (New York: Mentor, 1951).
23. I. S. Shklovskii and Carl Sagan: *Intelligent Life in the Universe* (New York: Dell, 1966).
24. Carl Sagan (ed.): *Communication with Extraterrestrial Intelligence* (Cambridge: M.I.T. Press, 1973).
25. Frank D. Drake: "How Can We Detect Radio Transmissions from Distant Planetary Systems?" and "Project Ozma," in Cameron: *Op. Cit.,* pp. 165–177.
26. Carl Sagan and Frank Drake: "The Search for Extraterrestrial Intelligence," *Scientific American,* May 1975, p. 84.
27. B. M. Oliver: "Some Potentialities of Optical Masers," in Cameron: *Op. Cit.,* pp. 207–222.
28. A. T. Lawton: "CETI from Copernicus," *Spaceflight,* August–September 1975, pp. 328–330.
29. "CETI Signal to Messier 13," *Spaceflight,* February 1975, pp. 54, 76. L. M. LaLonde: "The Upgraded Arecibo Observatory," *Science,* 18 October 1974, pp. 213–217.
30. Carl Sagan, Linda Sagan, and Frank Drake: "A Message from Earth," *Science,* 25 February 1972, pp. 881–884.
31. TRW Corporation: contest advertisement, *New Scientist,* 20 September 1973, p. 675.
32. Carl Sagan: *The Cosmic Connection* (New York: Dell, 1975) p. 33.
33. Sagan and Drake: *Op. Cit.,* p. 80.
34. *Ibid.,* p. 83.
35. *Ibid.*
36. John Lear: "The Search for Man's Relatives among the Stars," *Saturday Review,* 10 June 1972, pp. 30–37.
37. *The Gallup Opinion Index,* January 1974, pp. 20–23.
38. "A Man Whose Time Has Come," *New Scientist,* 4 July 1974, pp. 36–37.

REFERENCES

Agel, Jerome (ed.), *The Making of Kubrick's "2001"*, New American Library, New York, 1970.

Aldiss, Brian W., *Billion Year Spree,* Doubleday, Garden City, N.Y., 1973.

Aldrin, Edwin E. and Wayne Warga, *Return to Earth,* Random House, New York, 1973.

Anderson, Poul, *Is There Life on Other Worlds?,* Collier, New York, 1963.

Anonymous, U.S. Army, *The Story of Peenemünde* (intelligence documents), unpublished, 1945.

Armstrong, Neil et al., *First on the Moon,* Little, Brown, Boston, 1970.

Atheling, William, Jr. (James Blish), *The Issue at Hand,* Advent, Chicago, 1964.

———*More Issues at Hand,* Advent, Chicago, 1970.

Barber, Bernard, *Science and the Social Order,* Collier, New York, 1970 (1952).

Bauer, Raymond A., *Second-Order Consequences–A Methodological Essay on the Impact of Technology,* M.I.T. Press, Cambridge, 1969.

Baxter, John, *Science Fiction in the Cinema,* Paperback Library, New York, 1970.

Bergaust, Erik, *Reaching for the Stars,* Doubleday, Garden City, N.Y., 1960.

———and William Beller, *Satellite!,* Hanover House, Garden City, N.Y., 1956.

Berger, Peter L., *Invitation to Sociology,* Doubleday, Garden City, N.Y., 1963.

———*The Sacred Canopy,* Doubleday, Garden City, N.Y., 1969.

———and Thomas Luckmann, *The Social Construction of Reality,* Doubleday, Garden City, N.Y., 1967.

Bergson, Henri, *Creative Evolution,* Holt, New York, 1911.

Bhaktivedanta Swami, A. C., *Easy Journey to Other Planets,* ISKCON Press, Boston, 1970.

Bloomfield, Lincoln P. (ed.), *Outer Space–Prospects for Man and Society,* Praeger, New York, 1968.

Bonestell, Chesley and Willy Ley, *The Conquest of Space,* Viking, New York, 1951.

Bottome, Edgar M., *The Missile Gap,* Fairleigh Dickinson University Press, Rutherford, N.J., 1971.

Braun, Magnus Freiherr von, *Weg durch vier Zeitepochen,* Starke, Limburg an der Lahn, 1965.

Braun, Wernher von, *The Mars Project,* University of Illinois Press, Urbana, 1962.

———*Space Frontier,* Fawcett, Greenwich, Conn., 1969.

———and Frederick I. Ordway, *History of Rocketry and Space Travel,* Crowell, New York, 1971(?).

Briney, Robert E. and Edward Wood, *SF Bibliographies,* Advent, Chicago, 1972.

Brodie, Bernard and Fawn M. Brodie, *From Crossbow to H-Bomb,* Indiana University Press, Bloomington, 1973.

Cameron, A. G. W. (ed.), *Interstellar Communication,* Benjamin, New York, 1963.

Cantrill, Hadley, *The Invasion from Mars,* Harper & Row, New York, 1940.

Carpenter, M. Scott et al., *We Seven,* Simon and Schuster, New York, 1962.

Chapman, John L., *Atlas–The Story of a Missile,* Harper, New York, 1960.

Childe, V. Gordon, *Man Makes Himself,* Mentor, New York, 1951.

Chrysler, C. Donald and Don L. Chaffee, *...On Course to the Stars, the Roger B. Chaffee Story,* Kregel, Grand Rapids, Mich., 1968.

Clareson, Thomas D. (ed.), *SF: The Other Side of Realism,* Bowling Green University Popular Press, Bowling Green, O., 1971.

Clark, John D., *Ignition!–An Informal History of Liquid Rocket Propellants,* Rutgers University Press, New Brunswick, N.J., 1972.

Clarke, Arthur C. (ed.), *The Coming of the Space Age,* Meredith, New York, 1967.

———*The Exploration of Space,* Fawcett, Greenwich, Conn., 1964.

———*Profiles of the Future,* Bantam, New York, 1963.

———*Voices from the Sky,* Harper, New York, 1965.

———*2001–A Space Odyssey,* New American Library, New York, 1968.

Cleator, P. E., *Rockets through Space,* Simon and Schuster, New York, 1936.

Cole, Dandridge M., *Beyond Tomorrow,* Amherst Press, Amherst, Wisc., 1965.

Cookson, John and Judith Nottingham, *A Survey of Chemical and Biological Warfare,* Monthly Review Press, New York, 1969.

Cox, Donald and Michael Stoiko, *Spacepower,* Winston, Philadelphia, 1958.

Daniloff, Nicholas, *The Kremlin and the Cosmos,* Knopf, New York, 1972.

Davenport, Basil (ed.), *The Science Fiction Novel,* Advent, Chicago, 1969.

Day, Donald B., *Index to the Science-Fiction Magazines, 1926–1950,* Perri Press, Portland, Ore., 1952.

Dornberger, Walter, *V2–der Schuss ins Weltall,* Bechtle, Esslingen, Germany, 1952.

Durkheim, Emile, *Suicide,* Free Press, New York, 1951.

Ellik, Ron and Bill Evans, *The Universes of E. E. Smith,* Advent, Chicago, 1966.

Emme, Eugene M., *A History of Space Flight,* Holt, New York, 1965.

———*Hitler's Blitzbomber,* Air University, Maxwell Air Force Base, Ala., 1951.

Eney, Dick (ed.), *The Proceedings; Discon,* Advent, Chicago, 1965.

Erdmann, Herbert, *Dritter in Weltraum–Raumfahrttechnik in europäischen Ländern,* Schwann, Düsseldorf, 1969.

Essers, I., *Max Valier, ein Vorkämpfer der Weltraumfahrt, 1895–1930,* Technikgeschichte in Einzeldarstellungen, Verein Deutscher Ingenieure, 1968.

Etzioni, Amitai, *The Moon-Doggle,* Doubleday, Garden City, N.Y., 1964.

Evans, I. O., *Jules Verne and His Work,* Arco, London, 1965.

Ford, Brian, *German Secret Weapons,* Ballantine, New York, 1969.

Frankl, Viktor E., *Man's Search for Meaning,* Washington Square, New York, 1963.

Fritz, Alfred, *Start in die Dritte Dimension,* Harold, Stuttgart, 1958.

Fuller, R. Buckminster, *Operating Manual for Spaceship Earth,* Pocket Books, New York, 1970.

Gardner, Charles, *The Space Shuttle,* British Interplanetary Society, London, 1972.

Gardner, Martin, *Fads & Fallacies in the Name of Science,* Dover, New York, 1957.

Gartmann, Heinz, *The Men Behind the Space Rockets,* McKay, New York, 1956.

Gatland, Kenneth W. (ed.), *Project Satellite,* British Book Centre, New York, 1958.

Gerrold, David, *The World of Star Trek,* Ballantine, New York, 1973.

Gibney, Frank and George J. Feldman, *The Reluctant Space-Farers,* New American Library, New York, 1965.

Gilfillan, S. C., *The Sociology of Invention,* M.I.T. Press, Cambridge, 1963.

Goddard, Esther C. and G. Edward Pendray (eds.), *The Papers of Robert H. Goddard,* Vol. I, McGraw-Hill, New York, 1970.

Goddard, Robert H., *Autobiography,* St. Onge, Worcester, Mass., 1966.

———*Rocket Development,* Prentice-Hall, Englewood Cliffs, N.J., 1961.

Goldsen, Joseph M. (ed.), *Outer Space in World Politics,* Praeger, New York, 1963.

Goodrum, John C., *Wernher von Braun, Space Pioneer,* Strode, 1969.

Goodwin, Harold Leland, *The Images of Space,* Holt, New York, 1965.

———*Space: Frontier Unlimited,* Van Nostrand, Princeton, N.J., 1962.

Goulart, Ron, *An Informal History of the Pulp Magazines,* Ace, New York, 1972.

Green, Constance McLaughlin and Milton Lomask, *Vanguard: A History,* Smithsonian Institution, Washington, D.C., 1971.

Grey, Jerry and Vivian Grey (eds.), *American Rocket Society Space Flight Report to the Nation,* Basic Books, New York, 1962.

Grimwood, James M. et al., *Project Gemini–A Chronology,* NASA SP-4002, Washington, D.C., 1969.

Grissom, Virgil, *Gemini!* Macmillan, New York, 1968.

Gröttrup, Irmgard, *Die Besessenen und die Mächtigen,* Steingrüben, Stuttgart, 1958.

Gruber, William H. and Donald G. Marquis (eds.), *Factors in the Transfer of Technology,* M.I.T. Press, Cambridge, 1969.

Hahn, Fritz, *Deutsche Geheimwaffen, 1939–1945,* Vol. I, Hoffmann, Heidenheim, Germany, 1963.

Halberstam, David, *The Best and the Brightest,* Fawcett, Greenwich, Conn., 1973.

Haley, Andrew G., *Rocketry and Space Exploration,* Van Nostrand, Princeton, N.J., 1958.

Hansel, C. E. M., *ESP–A Scientific Evaluation,* Scribner's, New York, 1966.

Hartl, Hans, *Hermann Oberth–Vorkämpfer der Weltraumfahrt,* Oppermann, Hannover, W. Germany, 1958.

Hartt, Julian, *The Mighty Thor,* Duell, Sloan, and Pearce, New York, 1961.

Heiber, Helmut, *Goebbels,* Hawthorn, New York, 1972.

———(ed.), *Goebbels-Reden,* Droste, Düsseldorf, 1972.

Heinkel, Ernst, *Stürmisches Leben,* Mundus, Stuttgart, Germany, 1953.

Heisenberg, Werner, *Der Teil und das Ganze,* Piper, München, 1969.

Heuer, Kenneth, *Men of Other Planets,* Pelligrini & Cudahy, New York, 1951.

Hitch, Charles J. and Roland N. McKean, *The Economics of Defense in the Nuclear Age,* Atheneum, New York, 1965.

Holmes, David C., *The Search for Life on Other Worlds,* Bantam, New York, 1967.

Holzmann, Richard T., *Chemical Rockets,* Marcel Dekker, New York, 1969.

Hortleder, Gerd, *Das Gesellschaftsbild des Ingenieurs,* Suhrkamp, Frankfort, 1970.

———*Ingenieure in der Industriegesellschaft,* Suhrkamp, Frankfort, 1973.

Hoyle, Fred, *Of Men and Galaxies,* University of Washington Press, Seattle, 1964.

Hubbard, Earl, *The Search Is On,* Pace, Los Angeles, 1969.

Hubbard, L. Ron, *Dianetics–The Modern Science of Mental Health,* Paperback Library, New York, 1950.

Huzel, Dieter K., *Peenemünde to Canaveral,* Prentice-Hall, Englewood Cliffs, N.J., 1962.

Huzel, Dieter K. and David H. Huang, *Design of Liquid Propellant Rocket Engines,* NASA, Washington, D.C., 1971.

Hyman, Charles J., *German-English/English-German Astronautics Dictionary,* Consultants Bureau, New York, 1968.

Irving, David, *The German Atomic Bomb,* Simon and Schuster, New York, 1967.

———*The Mare's Nest,* Little, Brown & Co., Boston, 1964.

———(and Feldmarschall Milsch), *Die Tragödie der Deutschen Luftwaffe,* Ullstein, Frankfort, 1970.

Irwin, James B., *To Rule the Night,* Ballantine, New York, 1973.

Jacobs, Horace and Eunice Engelke Whitney, *Missile and Space Projects Guide,* Plenum, New York, 1962.

Jewkes, John, David Sawers, and Richard Stillerman, *The Sources of Invention,* Norton, New York, 1969.

Johnston, Lucile, *The Space Secret of the Universe,* Roberts & Sons, Birmingham, 1969.

Jones, H. Spencer, *Life on Other Worlds,* Mentor, New York, 1951.

Kahn, Herman, *On Escalation,* Penguin, Baltimore, 1968.

———and Anthony J. Wiener, *The Year 2000,* Macmillan, New York, 1967.

Kelen, Emery, *Stamps Tell the Story of Space Travel,* Thomas Nelson, Nashville, 1972.

Kemp, Earl (ed.), *The Proceedings; Chicon III,* Chicago, 1963.

Killen, John, *A History of the Luftwaffe 1915–1945,* Berkley, New York, 1967.

Klass, Philip J., *Secret Sentries in Space,* Random House, New York, 1971.

Klee, Ernst and Otto Merk, *Damals in Peenemünde,* Gerhard Stalling Verlag, Oldenburg and Hamburg, 1963.

Knight, Damon, *In Search of Wonder,* Advent, Chicago, 1967.

Kosmodemyansky, A., *Konstantin Tsiolkovsky—His Life and Work,* Foreign Languages Publishing House, Moscow, 1956.

Krieger, F. J. (ed.), *Behind the Sputniks,* Public Affairs Press, Washington, D.C., 1958.

Kuhn, Thomas S., *The Copernican Revolution,* Vintage, New York, 1959.

Langer, Walter C., *The Mind of Adolf Hitler,* Basic Books, New York, 1972.

Lasby, Clarence G., *Project Paperclip–German Scientists and the Cold War,* Atheneum, New York, 1971.

Lasswitz, Kurd, *Two Planets,* Southern Illinois University Press, Carbondale, 1971.

Lebedev, Vladimir and Yuri Gagarin, *Survival in Space,* Praeger, New York, 1969.

Lee, Asher, *Goering, Air Leader,* Hippocrene, New York, 1972.

———(ed.), *The Soviet Air and Rocket Forces,* Praeger, New York, 1959.

Lehman, Milton, *This High Man–The Life of Robert H. Goddard,* Pyramid, New York, 1963.

Leonard, Jonathan Norton, *Flight into Space,* Random House, New York, 1957.

Leonov, Alexei and Vladimir Lebedev, *Space and Time Perception by the Cosmonaut,* Mir, Moscow, 1971.

Levy, Lillian (ed.), *Space: Its Impact on Man and Society,* Norton, New York, 1965.

Lewis, C. S., *Out of the Silent Planet,* Macmillan, New York, 1965.

———*Perelandra,* Macmillan, New York, 1965.

———*That Hideous Strength,* Macmillan, New York, 1965.

Lewis, Richard S., *Appointment on the Moon,* Ballantine, New York, 1969.

Ley, Willy, *Rockets, Missiles, and Men in Space,* Signet, New York, 1969.

Lilly, John Cunningham, *The Mind of the Dolphin: A Nonhuman Intelligence,* Avon, New York, 1967.

Lindaman, Edward B., *Space: A New Direction for Mankind,* Harper, New York, 1969.

Logsdon, John M., *The Decision to Go to the Moon,* M.I.T. Press, Cambridge, 1970.

Lowell, Percival, *Mars as an Abode of Life,* Macmillan, New York, 1908.

Lunan, Duncan, *Interstellar Contact,* Regnery, Chicago, 1974.

Lundwall, Sam J., *Science Fiction: What It's All About,* Ace, New York, 1971.

Lupoff, Richard A., *Edgar Rice Burroughs: Master of Adventure,* Ace, New York, 1965.

Lusar, Rudolf, *Die deutschen Waffen und Geheimwaffen des 2. Weltkrieges und ihre Weiterentwicklung,* Lehmanns, München, 1971.

McGolrick, Joseph E. (ed.), *Launch Vehicle Estimating Factors for Advance Mission Planning,* NASA, Washington, D.C., 1972.

McGovern, James, *Crossbow and Overcast,* Hutchinson, London, 1964.

MacGowan, Roger A. and Frederick I. Ordway, *Intelligence in the Universe,* Prentice-Hall, Englewood Cliffs, N.J., 1966.

McPhee, John, *The Curve of Binding Energy,* Ballantine, New York, 1975.

Macvey, John W., *How We Will Reach the Stars,* Macmillan, New York, 1969.

Mailer, Norman, *Of a Fire on the Moon,* New American Library, New York, 1971.

Maltin, Leonard (ed.), *TV Movies,* Signet, New York, 1969.

Maslow, Abraham H., *Religions, Values and Peak-Experiences,* Viking, New York, 1970.

Mazlish, Bruce (ed.), *The Railroad and the Space Program,* M.I.T. Press, Cambridge, 1965.

Medaris, John B., *Countdown for Decision,* Putnam, New York, 1960.

Merton, Robert K., *The Sociology of Science,* University of Chicago Press, 1973.

Mielke, Heinz, *Der Weg zum Mond,* Verlag Neues Leben, Berlin, DDR, 1969.

Moore, Patrick, *Moon Flight Atlas,* Rand McNally, Chicago, 1969.

Moskowitz, Sam, *Explorers of the Infinite,* World, Cleveland, 1963.

———*The Immortal Storm,* Hyperion, Westport, Conn., 1974.

———*Seekers of Tomorrow,* Hyperion, Westport, Conn., 1974.

Nayler, J. L., *A Dictionary of Astronautics,* Hart, New York, 1964.

Noordung, Hermann, *Das Problem der Befahrung des Weltraums,* Schmidt, Berlin, 1928.

Oberth, Hermann, *Man into Space,* Harper, New York, 1957.

———*Stoff und Leben,* Reichl, Remagen, Germany, 1959.

———*Die Rakete zu den Planetenräumen,* Oldenbourg, München, 1923.

Odishaw, Hugh (ed.), *The Challenges of Space,* University of Chicago Press, Chicago, 1962.

Ofshe, Richard (ed.), *The Sociology of the Possible,* Prentice-Hall, Englewood Cliffs, N.J., 1970.

Ohring, George, *Weather on the Planets,* Doubleday, Garden City, N.Y., 1966.

Ordway, Frederick I., Carsbie C. Adams, and Mitchell R. Sharpe, *Dividends from Space,* Crowell, New York, 1971.

Ovenden, Michael W., *Life in the Universe,* Doubleday, Garden City, N.Y., 1962.

Overstreet, Robert M., *The Comic Book Price Guide–1974,* Overstreet, Cleveland, Tenn., 1974.

Panshin, Alexei, *Heinlein in Dimension,* Advent, Chicago, 1968.

Pardoe, G. K. C., *The Challenge of Space,* Chatto & Windus, London, 1964.

Parry, Albert, *Russia's Rockets and Missiles,* Doubleday, Garden City, N.Y., 1960.

Pendray, G. Edward, *The Coming Age of Rocket Power,* Harper, New York, 1945.

Peterson, Theodore, *Magazines in the Twentieth Century,* University of Illinois Press, Urbana, 1956.

Pocock, Rowland F., *German Guided Missiles,* Arco, New York, 1967.

Price, Alfred, *Luftwaffe,* Ballantine, New York, 1969.

von Puttkamer, Jesco, *Raum Stationen,* Verlag Chemie, Weinheim/Bergstr., Germany, 1971.

Rabinowitch, Eugene and Richard S. Lewis, *Man on the Moon–The Impact on Science, Technology, and International Cooperation,* Harper, New York, 1969.

Reitsch, Hanna, *Flying Is My Life,* Putnam's, New York, 1954.

———*Ich flog für Kwame Nkrumah,* J. F. Lehmanns, München, 1968.

Riabchikov, Evgeny, *Russians in Space,* Doubleday, Garden City, N.Y., 1971.

Rogers, Alva, *A Requiem for Astounding,* Advent, Chicago, 1964.

Ruland, Bernd, *Wernher von Braun–Mein Leben für die Raumfahrt,* Burda, Offenburg, Germany, 1969.

Ruzic, Neil P., *Where the Winds Sleep,* Doubleday, Garden City, N.Y., 1970.

Ryan, Cornelius (ed.), *Across the Space Frontier,* Viking, New York, 1952.

Rynin, N. A., *Interplanetary Flight and Communication,* 9 Volumes, Israel Program for Scientific Translations, Jerusalem, 1971.

Sagan, Carl (ed.), *Communication with Extraterrestrial Intelligence,* M.I.T. Press, Cambridge, 1973.

———*The Cosmic Connection,* Dell, New York, 1973.

Sänger, Eugen, *Raumfahrt–technische Überwindung des Krieges,* Rowohlt, Hamburg, Germany, 1958.

Sänger, Eugen and Irene Bredt, *A Rocket Drive for Long Range Bombers,* Technical Information Branch, Navy Department, 1952 (1944).

Sapolsky, Harvey M., *The Polaris System Development,* Harvard University Press, Cambridge, 1972.

Scarboro, C. W. (ed.), *20 Years of Space, the Story of America's Spaceport,* Scarboro, Cape Canaveral, Fla., 1969.

Schelling, Thomas C., *The Strategy of Conflict,* Oxford University Press, London, 1960.

Schmookler, Jacob, *Invention and Economic Growth,* Harvard University Press, Cambridge, 1966.

Shapley, Harlow, *The View from a Distant Star,* Dell, New York, 1963.

Sheldon, Charles S., *Review of the Soviet Space Program,* McGraw-Hill, New York, 1968.

Shklovskii, I. S. and Carl Sagan, *Intelligent Life in the Universe,* Dell, New York, 1966.

Shternfeld, Ari(o) *Soviet Space Science,* Basic Books, New York, 1959.

Simmel, Georg, *The Sociology of Georg Simmel,* Free Press, New York, 1950.

Simon, General Leslie E., *Secret Weapons of the Third Reich–German Research in World War II,* WE, Inc., Old Greenwich, Conn., 1971.

Smaus, Jewel Spangler and Charles B. Spangler, *America's First Spaceman,* Doubleday, Garden City, N.Y., 1962.

Smelser, Neil J., *Theory of Collective Behavior,* Free Press, New York, 1962.

Sobel, Lester A. (ed.), *Space: From Sputnik to Gemini,* Facts on File, New York, 1965.

Sorokin, Pitirim A., *The Ways and Power of Love,* Regnery, Chicago, 1967.

Sparks, James C., *Winged Rocketry,* Dodd, Mead, New York, 1968.

Speer, Albert, *Erinnerungen,* Propyläen, Berlin, 1969.

Stevens, Clifford J., *Astrotheology,* Divine Word, Techny, Ill., 1969.

Stoiko, Michael, *Soviet Rocketry: Past, Present, and Future,* Holt, New York, 1970.

Strauss, Erwin S., *Index to the SF Magazines, 1951–1965,* M.I.T. Science Fiction Society, Cambridge, 1966.

Sullivan, Walter, *We Are Not Alone,* New American Library, New York, 1966.

Swenson, Loyd S. et al., *This New Ocean–A History of Project Mercury,* NASA SP-4201, Washington, D.C., 1966.

Tantum, W. H. and E. J. Hoffschmidt (eds.), *The Rise and Fall of the German Air Force,* WE, Inc., Old Greenwich, Conn., 1969.

Teilhard de Chardin, Pierre, *The Future of Man,* Harper, New York, 1964.

Thomas, Shirley, *Men of Space,* Chilton, Philadelphia, 8 Volumes, 1960–1968.

Tokaev, G. A. (aka, Tokaty-Tokaev), *Stalin Means War,* Weidenfeld, London, 1951.

Tsiolkovsky, Konstantin, *The Call of the Cosmos,* Foreign Languages Publishing House, Moscow, 1960.

Tuck, Donald H., *The Encyclopedia of Science Fiction and Fantasy,* Advent, Chicago, 1974, Volume I.

U.S. News & World Report, *U.S. on the Moon,* Collier Books, Washington, D.C., 1969.

Vaeth, J. Gordon, *200 Miles Up,* Ronald, New York, 1956.

Van Dyke, Vernon, *Pride and Power–The Rationale of the Space Program,* University of Illinois Press, Urbana, 1964.

Viereck, Peter, *Meta-Politics: The Roots of the Nazi Mind,* Capricorn, New York, 1965.

Vladimirov, Leonid, *The Russian Space Bluff,* Dial, New York, 1973.

Waltz, George H., Jr., *Jules Verne,* Holt, New York, 1943.

Warner, Harry, Jr., *All Our Yesterdays,* Advent, Chicago, 1969.

Weinberg, Gerhard L. (ed.) *Hitlers Zweites Buch,* Deutsche Verlags-Anstalt, Stuttgart, 1961.

Wertham, Frederic, *The World of Fanzines,* Southern Illinois University Press, Carbondale, 1973.

Whitfield, Stephen E. and Gene Roddenberry, *The Making of Star Trek,* Ballantine, New York, 1968.

Wilford, John Noble, *We Reach the Moon,* Norton, New York, 1971.

Young, Hugo, Bryan Silcock, and Peter Dunn, *Journey to Tranquility,* Doubleday, Garden City, N.Y., 1970.

Young, Richard S., *Extraterrestrial Biology,* Holt, New York, 1966.

Zaehringer, Alfred J., *Soviet Space Technology,* Harper, New York, 1961.

INDEX